INTRODUCTION TO ELECTRIC POWER ENGINEERING

Richard D. Shultz
Clarkson University

Richard A. Smith
Florida Power Corporation

HARPER & ROW, PUBLISHERS, New York
Cambridge, Philadelphia, San Francisco,
London, Mexico City, São Paulo, Singapore, Sydney

Sponsoring Editor: John Willig
Project Editor: Cynthia L. Indriso
Cover Design: Lawrence R. Didona
Text Art: Reproduction Drawings, Ltd.
Production: Marion A. Palen
Compositor: Syntax International Pte. Ltd.
Printer and Binder: Maple Press

INTRODUCTION TO ELECTRIC POWER ENGINEERING
Copyright © 1985 by Harper & Row, Publishers, Inc.

All rights reserved. Printed in the United States of America. No part of this book may be used or reproduced in any manner whatsoever without written permission, except in the case of brief quotations embodied in critical articles and reviews. For information address Harper & Row, Publishers, Inc., 10 East 53d Street, New York, NY 10022.

Library of Congress Cataloging in Publication Data

Shultz, Richard D.
   Introduction to electric power engineering.

   Bibliography: p.
   Includes index.
   1. Electric engineering. I. Smith, Richard A.
(Richard Allan), 1951– . II. Title.
TK145.S555   1985   621.31   84-4554
ISBN 0-06-046131-4

85 86 87 88 9 8 7 6 5 4 3 2 1

Contents

PREFACE xi

Chapter 1 THE POWER SYSTEM 1

1.1 Introduction 1
1.2 Elements of a Power System 3
1.3 Load Characteristics 3
    1.3.1 Electric Load Growth 3
    1.3.2 Load Profiles 4
1.4 Economic Dispatch 5
1.5 Protection 5
1.6 Engineers, the Power Industry, and this Text 6
1.7 Notation and Symbols 6

Chapter 2 ELECTROMECHANICS 8

2.1 Introduction 8
2.2 Magnetic and Electric Fields 13
2.3 Maxwell's Equations 14
2.4 Quasi-Static-Field Approximation 15
2.5 Applications of Maxwell's Equations to Magnetic Circuits 16
2.6 Magnetic Field Energy 36
2.7 Force Calculation from the Expression for Magnetic Field Energy 39
2.8 Force and Coenergy Relationships 41
2.9 Summary 44
2.10 Problems 44

Chapter 3 SYNCHRONOUS GENERATORS 48

3.1 Introduction 48
3.2 Voltage Induced in Synchronous Generators 52
    3.2.1 Elementary Synchronous Generators 52
    3.2.2 Voltage Induced in a Synchronous Generator 54
3.3 Torque in Synchronous Generators 59
    3.3.1 Flux Linkages in a Synchronous Generator 60

3.3.2 Calculation of Magnetic Field Coenergy and τ^e 66
3.3.3 Physical Interpretation of τ^e and γ 68
3.4 Multipole Synchronous Generators 71
3.5 Equivalent Electric Circuit of a Synchronous Machine 72
3.6 Power and Torque Angle 78
3.7 Synchronous Generators in a Power System 80
3.8 Summary 85
3.9 Problems 85

Chapter 4 POWER TRANSFORMERS 88

4.1 Introduction 88
4.2 Single-Phase Transformers 93
    4.2.1 Voltage Transformation 94
    4.2.2 Current Transformation 95
    4.2.3 Exciting Current 96
    4.2.4 Measuring r_c and X_m 98
    4.2.5 Referring Impedances 99
    4.2.6 Measuring Coil Resistance and Leakage Reactance 102
    4.2.7 Approximate Equivalent Circuits 103
    4.2.8 Physical Considerations 106
    4.2.9 Multiwinding Transformers 106
    4.2.10 Autotransformers 108
    4.2.11 Core Nonlinearities 109
4.3 Three-Phase Transformer Banks 111
    4.3.1 Y-Y Connection 111
    4.3.2 Δ-Δ Connection 112
    4.3.3 Δ-Y and Y-Δ Connections 114
    4.3.4 Three-Phase Transformers 116
4.4 Three-Phase Transformation with Two Transformers 117
4.5 Summary 118
4.6 Problems 118

Chapter 5 TRANSMISSION LINES 122

5.1 Introduction 122
5.2 Transmission Line Resistance 126
5.3 Transmission Line Inductance 127
    5.3.1 Internal Inductance 127
    5.3.2 Inductance Due to External Flux Linkage 129
    5.3.3 Inductance of Single-Phase Transmission Lines 131
    5.3.4 Inductance of Three-Phase Transmission Lines 132
    5.3.5 Geometric Mean Radius of Stranded Conductors 135

5.4   Capacitance of Transmission Lines   137
    5.4.1   Capacitance Calculations for Single-Phase Transmission Lines   139
    5.4.2   Capacitance Calculations for Three-Phase Transmission Lines   142
5.5   Bundled Conductors in Three-Phase Transmission Lines   145
5.6   Electrical Network Models of Transmission Lines   146
5.7   Real- and Reactive-Power Flow on a Transmission Line   150
5.8   Summary   152
5.9   Problems   152

Chapter 6   NETWORK ANALYSIS   156

6.1   Introduction   156
6.2   One-Line Diagrams and Impedance Diagrams   156
6.3   Per Unit Quantities   158
    6.3.1   Changing per Unit Bases   164
    6.3.2   Determining per Unit Bases   165
    6.3.3   Per Unit Single-Phase Impedance Representation of Δ-Connected Devices   170
    6.3.4   The Advantages of per Unit Versus Ohmic Representation   173
6.4   Power Flow Analysis   173
    6.4.1   Bus Admittance Matrix   173
    6.4.2   Classification of Buses   176
    6.4.3   Load Bus Equations   177
    6.4.4   Solution of Nonlinear Equations   178
    6.4.5   Voltage Control Buses   181
    6.4.6   The Swing Bus   182
    6.4.7   Initial Voltage Estimates   182
    6.4.8   Results Obtained from a Load Flow Analysis   184
6.5   Control of Power Flow   184
    6.5.1   Synchronous Generators   184
    6.5.2   Capacitor Banks   187
    6.5.3   Tap-Changing Transformers   187
6.6   Summary   190
6.7   Problems   190

Chapter 7   ELECTRIC MOTORS   193

7.1   Introduction   193
7.2   Elementary Motor Operation   193
7.3   Synchronous Motors   197
7.4   Three-Phase Induction Motors   200
    7.4.1   Operation of the Three-Phase Induction Motor   200
    7.4.2   Rotor Quantities at Standstill   201
    7.4.3   Running Rotor   202

   7.4.4 Equivalent Circuit of the Induction Motor 205
   7.4.5 Analyzing the Equivalent Circuit 209
   7.4.6 The Equivalent Circuit at Start-up 213
   7.4.7 Determining Power and Torque from the Thevenin Equivalent 214
   7.4.8 Calculation of Maximum Torque 216
   7.4.9 Varying Rotor Resistance and Starting Torque 218
 7.5 Single-Phase Induction Motors 219
   7.5.1 Magnetic Field of the Stator 220
   7.5.2 Capacitive Starting Single-Phase Induction Motors 222
   7.5.3 Split-Phase Single-Phase Induction Motors 223
   7.5.4 Equivalent Circuit of a Rotating Single-Phase Motor 223
 7.6 Direct-Current Motors 228
   7.6.1 Construction of dc Motors 229
   7.6.2 Voltage Induced on the Armature of a dc Motor 232
   7.6.3 Power and Torque in dc Motors 234
   7.6.4 Field and Armature-Winding Connections 236
 7.7 Universal Motors 238
 7.8 Summary 239
 7.9 Problems 240

Appendix A PHASOR ANALYSIS 243

A.1 Phasor Representation of Sinusoids 243
A.2 Impedances 246

Appendix B AVERAGE POWER AND THREE-PHASE CALCULATIONS 248

B.1 Average Power 248
B.2 Power Factor and Complex Power 249
B.3 Three-Phase Calculations 251

Appendix C MATRIX ALGEBRA 255

C.1 Definitions 255
C.2 Algebraic Operations 255
C.3 Inverse of a Matrix 256
C.4 Matrix Partitioning 256

BIBLIOGRAPHY 258

INDEX 261

Preface

The objective of this text is to introduce to engineering students at an early level the principles and concepts which are the basis of the electric power industry. The material starts with the synchronous generator and continues with the step-up transformer, the transmission system, and the loads. Although this presentation is the method the authors prefer, each topic is treated individually and may be taught in any order. By presenting electric machinery as a part of the total power system, the authors have tried to avoid the traditional separation of electric machinery and power system analysis courses.

The material in this text is designed to be taught in a junior-level, one-semester course. It is assumed that the students have had one course covering basic circuit concepts.

Chapter 1 is an introductory chapter. It is a brief history and a description of the present role of energy in society. The manner in which the electric power engineer (and this text) fits into that role is also discussed.

Chapter 2 familiarizes students with electromechanical concepts, quantities, and calculation procedures that are encountered in subsequent chapters. Calculation techniques for quantities such as flux linkage, energy, and force in electromechanical devices are provided. The techniques assume quasi-static-field approximations along with restrictions on magnetic flux leakage, field fringing at air gaps, and magnetic permeability.

Chapter 3 begins the study of power systems with sychronous generators. The first part of the chapter develops an equivalent electric circuit model of a generator. The chapter continues with explanations of these machines by using the circuit model to calculate the electric power and torque developed in them.

In Chapter 4 the power transformer is discussed. The first part of the chapter presents the concepts of voltage and current transformation. Then the equivalent circuits of both single-phase and three-phase transformer banks are developed.

Chapter 5 continues the study of power systems by examining transmission lines. Part of the purpose of this chapter is to show how quantities such as resistance, inductance, and capacitance of transmission lines are calculated. These quantities are then used later in the chapter to develop a transmission line electric circuit model. The chapter concludes with illustrations of power flow calculations using the circuit model.

In Chapter 6 the concepts involved in power system analysis are discussed. These concepts include the per unit system, a power flow analysis, and methods of controlling power flow.

Chapter 7 concludes the text by examining electric loads. Since electric motors comprise the major portion of loads, this chapter confines its study to motors. Three-phase synchronous and induction motors are discussed first. This discussion is followed by a look at single-phase motors. The chapter concludes with a study of direct-current motors.

This text has two basic objectives. It is meant to prepare those students choosing power engineering as their area of expertise for more advanced courses. This is accomplished by showing them the fundamentals involved in power system operation and analysis by introducing one element at a time until the system is complete. The other objective of this text is to introduce those students not choosing power engineering as a profession to some of the basic principles of power system operation and energy conversion from a whole-system point of view.

The authors have received a great deal of assistance from many people in the preparation of this text. In particular, we would like to thank Professors Peter Sauer and Philip Krein of the University of Illinois for their many valuable suggestions. The editorial staff of Harper & Row has been very helpful and encouraging. Finally, we acknowledge our great debt to Ms. Emma Haynes for her skill and patience in typing this manuscript.

Richard D. Shultz
Richard A. Smith

Chapter 1

The Power System

1.1 INTRODUCTION

One of the primary contributors to the advancements and improvements in man's life-style over the years has been his ability to use and control energy. Man's use of energy can be seen in everyday operations such as mechanical motion and the production of heat and light. As civilization advanced, the consumption of energy in homes and factories in many parts of the world increased beyond the point where useful forms of energy could be economically and safely produced at the same location it was being used. The emergence of large central energy generating/processing stations with elaborate transmission and distribution systems were the result. Oil processing facilities, steam generating plants, and electric generating stations are all a part of our total energy system.

One of the most economical, easiest, and safest ways to transmit energy is in the form of electric energy. Energy is converted at the generating station from its basic source (fossil fuels, hydro, or nuclear) into electric energy. This electric energy is then sent over a transmission system to various loads where it is usually converted into other useful forms of energy. Electric energy is used primarily to transmit the energy from one source, such as heat from burning coal, at one location to another location to do work, such as running a compressor on a refrigerator. In this book we shall investigate how these conversions and transmissions of energy take place and study the devices used in the operation of an electric power system.

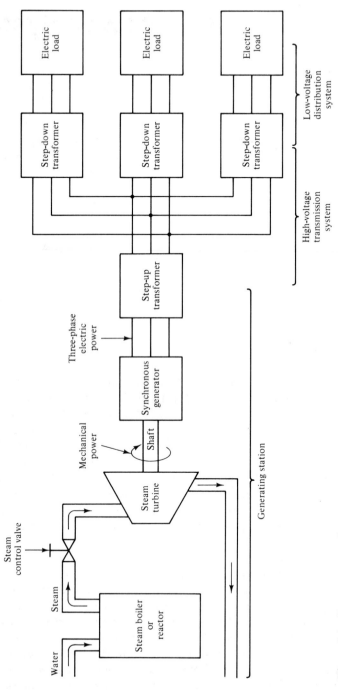

Figure 1.1 Power system elements.

1.2 ELEMENTS OF A POWER SYSTEM

This textbook is structured such that a power system is described element by element, chapter by chapter. The major elements, as they appear in a complete power system, are shown in Figure 1.1.

Starting on the left side of the diagram, the first major power system component is a generating station. The generating station has the function of converting energy from its basic form into electric energy. Sources of basic energy presently include fossil fuel (oil, natural gas, and coal), nuclear fuel, and water flow. Other forms of basic energy presently undergoing research and development include solar, wind, geothermal, and even garbage.

The most common use of basic energy is to convert water into steam within a boiler. The steam is used to drive a turbine, which rotates the shaft of a synchronous generator. The generator converts the mechanical power of the shaft into electric power, usually in the form of three-phase electricity.

The last element in the generating station that handles the three-phase power is the step-up transformer. The transformer increases the generator's low terminal voltage to a higher voltage. This increase in voltage is accompanied by a decrease in current on the high-voltage terminals of the transformer (power into the transformer equals power out of the transformer, which means that low voltage and high current into the transformer results in high voltage and low current out of the transformer).

The high-voltage side of the transformer is connected to the high-voltage transmission system. Figure 1.1 shows three transmission lines leaving the generating station. These lines carry high-voltage, low-current electric power in order to reduce their I^2r losses. Once the transmission lines have reached customer load centers, their voltages are reduced to customer usable levels through step-down transformers. The electric power leaving the transformers is carried to individual customer loads by the low-voltage distribution lines.

The electric loads connected to the distribution lines use electric power in a variety of ways. However, the majority of the power is used by electric motors. Other uses include lighting, heating, arc furnaces, rectifier loads, and so forth.

1.3 LOAD CHARACTERISTICS

The electric power system is designed to deliver electric energy efficiently and safely to the customer. The characteristics of electric energy demand sometimes make this task difficult. Predicting the rate of load growth and meeting the daily and yearly load cycles are two difficult challenges.

1.3.1 Electric Load Growth

From the 1940s until the early 1970s the annual growth rate of electric energy demand in this country was about 7 percent. In the late 1970s and early 1980s conservation methods and economic conditions slowed the growth rate to about 2 to 3 percent. Even at this slower rate, the demand for electric energy will double every 25 to 35 years. Depending on the type of power plant, a total of 8 to 10 years can elapse from the time of conception to completion of a large electric generating station. System planners must, therefore, study their power system 10 to 20 years in the future in order to have sufficient time to predict and correct future energy problems.

Through a combination of examining historic trends and making futuristic predictions, the planner estimates future generation requirements and recommends facility construction. The job of the system planner does not, however, consist only of ensuring sufficient generation for new loads. He must also determine

1. if existing transmission lines and equipment can adequately carry the additional energy requirements from the generating stations to the loads
2. if the system equipment is adequately protected under fault conditions for the new operating points
3. if any transient conditions will disrupt the normal operation of the system
4. the most economical operating state for various loading conditions.

In addition to these technical challenges, which arise with increasing loads, the problems of environmental impact and public acceptance of new facilities must also be addressed. In order to satisfy the ever-increasing electric energy needs of our society, electric power engineers must make these predictions and solve these problems and more on a continuous basis.

1.3.2 Load Profiles

Since technology has not yet provided a means for storing electric energy in an efficient and cost-effective manner, electric power must be generated as it is needed. The customer demand for power varies throughout the day. A typical demand versus time curve is shown in Figure 1.2. Starting a large electric power station requires a significant amount of time and is an expensive procedure. Thus, even though the large stations are usually more economical to operate,

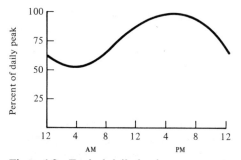

Figure 1.2 Typical daily load curve.

1.5 PROTECTION

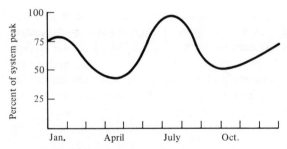

Figure 1.3 Typical yearly load curve.

if additional generation is needed for only a few hours to meet the peak load, smaller peaking generators would probably be used. These peaking units are less efficient and more expensive to operate but have lower start-up costs. This situation illustrates that not only must power engineers predict the growth of load over years in order to meet generation needs, but the hourly loads must also be estimated in order to economically meet each day's demand.

Just as the load varies each hour in the day, the peak load of each day varies throughout the year. Figure 1.3 shows a typical curve of daily peak loads versus time for 1 year of a summer-peaking utility.

The yearly variation in load is very important in scheduling maintenance of equipment. During the peak load periods, having all the equipment available on-line to meet the demands of the system is desirable. Therefore, the low-load periods of the curve in Figure 1.3 are the times when routine maintenance is usually scheduled. Taking equipment out of service for maintenance affects the economics of power production and decreases the safety margins between demand for and production capability of electric power. The power engineer must consider these effects, as well as the necessity for maintenance, in determining when to take equipment out of service.

1.4 ECONOMIC DISPATCH

In addition to predicting future energy demands, the power engineer must supply those demands in the most economical manner available to him. This is not an easy task, since fuel, labor, and maintenance costs vary between generating stations. Even at each generating station, the efficiency of the plant varies with the output of that plant. Taking all of these costs into account, the engineer must determine how to load each generator at each plant so as to minimize the cost of delivering power to the customer. This process is called economic dispatch.

1.5 PROTECTION

A fault is any condition that interferes with the normal flow of current in the power system. Faults can result from severe weather conditions, foreign objects short-circuiting one or more conductors, failure of insulation, or damage to structures that separate conductors from each other and ground. Faults can result in unacceptable voltage levels, abnormally large current flows, arcing, and the inability of the power system to operate as a whole. It is the responsibility of the engineer to design a protection scheme of fuses and circuit breakers which detects faults quickly and isolates them from the rest of the system. With such a scheme the fewest number of customers will be affected by any particular fault condition.

1.6 ENGINEERS, THE POWER INDUSTRY, AND THIS TEXT

This brief introduction shows that many challenges are waiting for the future electric power engineer. One of the objectives of this book is to help students choosing the area of power engineering as their career to prepare for further study and eventual employment in the industry. The power industry is different from most industries in that it is regulated by state and federal governments. So not only does the industry touch virtually everyone through the service it provides, but everyone has the ability to affect the industry through the regulatory bodies. Another objective of this book is to educate engineers who do not choose power engineering as a profession in some of the basic principles of power system operation. In this way they can exercise their input into the industry from an informed point of view.

This text is designed for a junior-level, one-semester or equivalent course. It is assumed that students at this level have an understanding of basic circuit concepts. After studying this text, the student will hopefully have a general idea of how the elements of a power system fit and work together. This concept should allow those students taking advanced power courses to understand how each specialized topic fits in the overall scheme of the system. In addition, this total power system approach will give those students not continuing in the power field a basic understanding of how an entire power system works in one text.

1.7 NOTATION AND SYMBOLS

An attempt has been made to keep the notation in this book as simple as possible. The use of small letters for ac quantities, such as i and v, indicates instantaneous values. The use of capital letters means that the values are phasor quantities with a magnitude and angle. The magnitude of phasors will be rms values unless otherwise stated (see Appendix A). If only the magnitude of the phasor quantity is desired, then the absolute value symbol $|I|$ is used. In some equations, where the answer is a real number with a magnitude only, the absolute value symbols have been deleted since the operation is obvious. For

1.7 NOTATION AND SYMBOLS

example, $P = VI\cos\theta$ results in a real number, so V and I are magnitudes only. The student is in most cases reminded of this fact in the text.

For three-phase calculations real, reactive, and complex powers are assumed to be three-phase quantities unless otherwise stated. Similarly, voltages are assumed to be line-to-line values unless stated otherwise.

These definitions and notations are explained in detail as they are used in the text. A complete understanding and appreciation of the notation will come with practice.

Chapter 2
Electromechanics

2.1 INTRODUCTION

The title of this chapter, "Electromechanics," emphasizes one important aspect of power systems; that is, the conversion of energy from mechanical to electrical form and from electrical to mechanical form. Both of these conversions take place in power systems and can be viewed in a simple block diagram as shown in Figure 2.1.

The block on the left represents the generation system. This system serves the purpose of converting mechanical energy into electric energy. In most cases the mechanical energy is in the form of a spinning shaft driven by a steam turbine, and the conversion of mechanical energy into electric energy takes place within rotating machines called synchronous generators.

The middle block of Figure 2.1 represents the part of a power system that carries the electric energy from the generation system to the load system. It consists of three-phase and single-phase transmission lines that are energized to voltages ranging from 120 V to 765 kV. The transmission network is planned and operated to minimize the electric energy lost in transferring energy from the generation system to the load system.

Figure 2.1 Block diagram of a power system.

2.1 INTRODUCTION

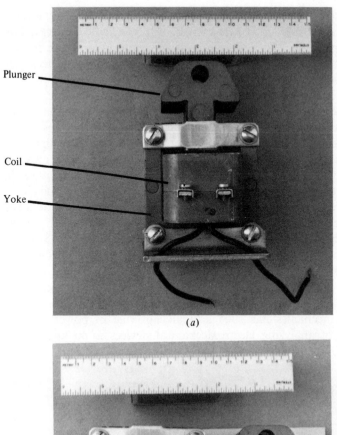

(a)

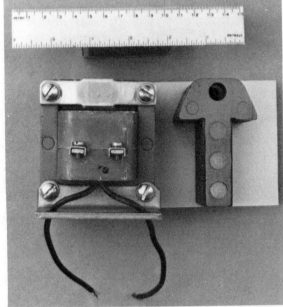

(b)

Figure 2.2 (a) Solenoid with plunger suspended in yoke. (b) Solenoid with plunger removed from yoke.

The load system is represented in Figure 2.1 by the block on the right. It shows that electric energy is in part converted to mechanical energy. This conversion takes place in rotating machines called motors. However, motors represent only part of the electric load in the load system. Other devices such as lights and heating elements use electric energy.

Figures 2.2 to 2.5 show examples of various kinds of electromechanical devices. Figure 2.2 shows a solenoid consisting of a movable plunger whose motion is controlled by a coil of wire. Application of current to the coil causes a force to be exerted on the plunger that pulls it into the iron yoke. Figure 2.3 shows a three-phase circuit breaker that has two operating coils. A movable plunger is positioned in both coils and is drawn left or right when either the left or right coil is energized. This left-right motion opens and closes the circuit connecting the three large terminals on the left and right sides of the breaker. Figure 2.4 shows a small shaded pole motor with several coils located around the periphery of the stator. Current flowing in these coils causes a torque to develop on the rotor, forcing it to rotate. Figure 2.5 shows a small universal motor. It has coils on both the rotor and stator. Torque is developed on the rotor when current is applied to the coils. Each of the electromagnetic devices shown is typical of the kinds of equipment found in power systems.

Since electromechanics has such a significant role in power systems, learning methods of analyzing electromechanical devices is an appropriate

Figure 2.3 Circuit breaker with two coils.

2.1 INTRODUCTION

(a)

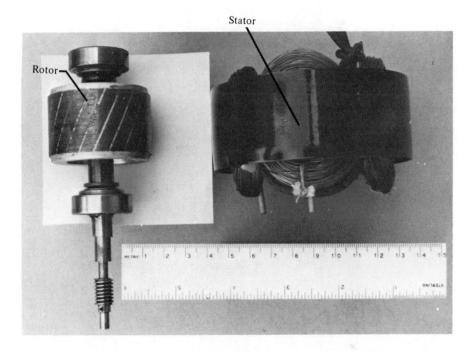

(b)

Figure 2.4 (a) Shaded pole motor with rotor suspended inside the stator. (b) Shaded pole motor with rotor removed from stator.

(a)

(b)

Figure 2.5 (a) Universal motor with rotor suspended inside stator. (b) Universal motor with rotor removed from stator.

place to begin the study of power systems. Therefore, the purpose of this chapter is to develop the techniques of analyzing electromechanical devices such as synchronous generators and electric motors.

2.2 MAGNETIC AND ELECTRIC FIELDS

The conversion of energy between mechanical and electrical forms is performed through magnetic fields. Magnetic fields are capable of causing both electric and mechanical phenomena. From an electrical point of view, magnetic fields are capable of inducing voltages in conductors. From a mechanical point of view, magnetic fields are capable of causing forces and torques of attraction and repulsion. Thus, the magnetic field in an electromechanical device is an important quantity to study.

A magnetic field is established by the motion of an electron. Since current in a wire is the flow of electrons, a current-carrying wire has a magnetic field established around it. The field has properties of direction, density, and intensity and is most easily portrayed as consisting of "lines of flux." The magnetic flux is referred to by the symbol

$$\phi \triangleq \text{magnetic flux with units of webers}$$

The property of direction of magnetic flux "flow" is related to current by the right-hand rule. If the thumb of the right hand is pointed in the direction of current flow in a wire, the fingers of the right hand will point in the direction of the magnetic flux flow caused by the current. Figure 2.6 illustrates the right-hand rule. The illustration shows that magnetic flux completely encircles the wire and closes on itself.

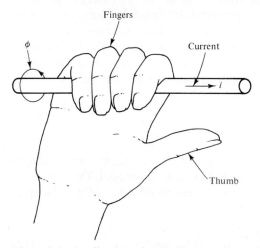

Figure 2.6 Application of right-hand rule to current-carrying wire.

The property of magnetic field density is defined by the number of flux lines piercing a known area. Flux density is referred to by the symbol

$$B \triangleq \frac{\text{number of flux lines piercing an area}}{\text{area}} \quad \text{Wb/m}^2$$

Note that B has units of webers per square meter (Wb/m^2) and is a vector with magnitude and direction. Its direction is the same as the direction for ϕ.

The property of magnetic field intensity is defined by flux magnitude along a known distance. It is referred to by the symbol

$$H \triangleq \frac{\text{flux magnitude along a distance}}{\text{distance}} \quad \text{A/m}$$

Magnetic field intensity H is also a vector quantity, with a direction the same as that of ϕ.

The properties of B and H are related to each other by the permeability of the medium in which a magnetic field has been established. The relationship is

$$B = \mu H$$

where $\mu \triangleq$ permeability, with units of henrys per meter (H/m).

The permeability of free space is referred to as μ_0 and has a value of $4\pi \times 10^{-7}$ H/m. Materials such as iron and nickel have relatively high permeabilities compared to free space. These kinds of materials are said to exhibit ferromagnetic characteristics. Their permeabilities range from several hundred to 1 million times larger than μ_0. Permeability is sometimes expressed as a function of μ_0 by defining a multiplying factor μ_r called relative permeability; that is, $\mu = \mu_r \mu_0$.

A magnetic field can induce voltages in an electric energy system. A voltage is usually thought of as a potential difference between two points. This difference is established by electric charges (i.e., electrons and positive ions) which have been separated by the action of a magnetic field. The existence of electric charges causes electric fields to exist. The electric field intensity E is one quantity that describes these fields. E has units of volts per meter.

2.3 MAXWELL'S EQUATIONS

In order to study the effects of fields in magnetic devices, relationships between electric fields, magnetic fields, currents, and charges are needed. These relationships have been derived and cast into a set of equations called Maxwell's equations. They appear in integral form as

$$\oint_l H \cdot dl = \int_s \left(J + \varepsilon \frac{\partial E}{\partial t} \right) \cdot ds \quad \text{(Ampere's law)} \quad (2.1)$$

$$\oint_{l_t} E' \cdot dl = \frac{-d}{dt} \int_s B \cdot ds \quad \text{(Faraday's law)} \tag{2.2}$$

$$\oint_s E \cdot ds = \int_v \frac{\rho}{\varepsilon} dV \quad \text{(Gauss' law)} \tag{2.3}$$

$$\oint_s B \cdot ds = 0 \quad \text{(magnetic source law)} \tag{2.4}$$

$$\oint_s J \cdot ds = \frac{-d}{dt} \int_v \rho \, dV \quad \text{(conservation of charge law)} \tag{2.5}$$

where J = current density, A/m²
 ρ = charge density, C/m³

Two notes are in order concerning the notation for electric field intensity: E' in Faraday's law and the applicability of Gauss' law to magnetic devices. The prime notation indicates that motion of the contour in a magnetic field is also considered; that is, E' is given as

$$E' = E + u \times B$$

where E = electric field intensity in a fixed reference frame
 u = velocity of the contour, m/s
 B = magnetic field intensity
 E' = electric field intensity in the moving reference frame associated with u

Thus, an electric field E can be induced along a contour not only by time variation of a magnetic field through the contour but also by motion of the contour through a constant or even time-varying magnetic field.

Gauss' law relates the charge contained in and the electric field radiating from a closed volume. Since this chapter is concerned with magnetic devices, the use of Gauss' law is not applicable and will not be considered.

2.4 QUASI-STATIC-FIELD APPROXIMATION

Maxwell's equations can be simplified for a large class of problems by applying quasi-static-field approximations. In these approximations the speed of propagation of electromagnetic fields (speed of light) is so fast relative to the short mechanical distances and low frequencies of excitation of the devices studied that variations of electric fields across the devices at any instant are negligible. This approximation has impact on magnetic field systems in Ampere's law and the conservation of charge law. The time variation of the electric field may be neglected as a source of magnetic field. Thus, Ampere's law appears as

$$\oint_{l_t} H \cdot dl = \int_s \left(J + \varepsilon \frac{\partial E}{\partial t}\!\!\!\!\!\!\!\!\!\nearrow^{0} \right) \cdot ds$$

$$= \int_s J \cdot ds \tag{2.6}$$

In the conservation of charge law the time rate of change of electric charge may be neglected. Thus, this law reduces to

$$\oint_s J \cdot ds = 0 \tag{2.7}$$

which is easily interpreted as Kirchhoff's current law.

The remaining equations of interest for magnetic field systems are

$$\oint_l E' \cdot dl = \frac{-d}{dt} \int_s B \cdot ds \quad \text{(Faraday's law)} \tag{2.8}$$

$$\oint_s B \cdot ds = 0 \quad \text{(magnetic source law)} \tag{2.9}$$

$$B = uH \tag{2.10}$$

2.5 APPLICATIONS OF MAXWELL'S EQUATIONS TO MAGNETIC CIRCUITS

Equations 2.6 to 2.10 can be used to analyze phenomena of interest in magnetic devices. One of the phenomenon of interest is the voltage induced by the excitation current applied to the devices. The following examples illustrate the methods needed to calculate voltage.

EXAMPLE 2.1

Consider the device shown in Figure 2.7. This device is a single-air-gap, iron core inductor. The voltage of interest is v_{coil}, the voltage induced on the coil of wire wrapped around the left leg. The inductor is constructed on an iron core with a square cross section of width W and depth D. It has a small air gap in its right leg of width g. The voltage, v_{coil}, is a function of current, number of turns of wire in the coil, and iron core dimensions.

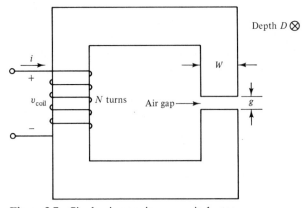

Figure 2.7 Single air gap, iron core inductor.

2.5 APPLICATIONS OF MAXWELL'S EQUATIONS TO MAGNETIC CIRCUITS

Examining Equations 2.6 to 2.10 again and assuming the current is known, Faraday's law looks like a good place to begin analyzing this device. This starting point is selected because electric field intensity E' appears on the left-hand side of Equation 2.8 with units of volts per meter. Since volts are of interest in this example, calculation of E' is necessary in order to find v_{coil}.

The voltage that appears across the coil is in part a function of the voltage at the terminals of each of the loops of wire in the coil. Consider the loop at the top of the coil, as shown in Figure 2.8. Current is entering the upper terminal of the loop and generating a magnetic field B, which, by the right-hand rule, has a direction out of the page. This direction is indicated by the circle with a dot in its center. A voltage v_{loop} is assumed to appear at the terminals, with polarity as indicated.

Application of Faraday's law to this loop requires the selection of a contour of interest. Since voltage is the quantity to be derived from the electric field, the contour of interest should be selected such that the electric field along the contour can be used to find v_{loop}. Therefore, the obvious choice for contour position is around the loop of wire and across its terminals. This contour is shown in Figure 2.8 as the vector dl. Its assumed direction specifies the direction of the surface vector ds by the right-hand rule. Thus, the vector ds is shown pointing out of the page.

Consider the left-hand side of Faraday's law (Equation 2.8) as applied to the conditions of Figure 2.4.

$$\oint_l E' \cdot dl = \int_1^2 E' \cdot dl + \int_2^1 E' \cdot dl \tag{2.11}$$

The first integral on the right-hand side of Equation 2.11 is the contour integration from the upper loop terminal 1 through the loop and back to the lower loop terminal 2. The second integral is the contour integration across the gap of the terminals from 2 to 1. Two assumptions are now made about the loop.

1. The loop of wire is a perfect conductor. This assumption means that no electric field exists inside the loop.

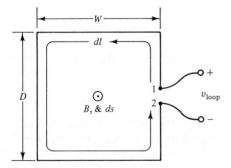

Figure 2.8 Looking down on top loop of coil.

2. The contour across the loop terminals is spatially oriented such that no time-varying magnetic field exists around it.

The first assumption reduces the first integral of Equation 2.11 to zero because its operand is zero. Equation 2.11 now appears as

$$\oint_{l_t} E' \cdot dl = \int_1^2 \overset{0}{\cancel{E'} \cdot dl} + \int_2^1 E' \cdot dl = \int_2^1 E' \cdot dl \qquad (2.12)$$

At this point $E + u \times B$ is substituted for E', and since $u = 0$, Equation 2.12 is further simplified to

$$\oint_{l_t} E' \cdot dl = \int_2^1 E \cdot dl \qquad (2.13)$$

The second assumption establishes static-field conditions across the terminals. These conditions mean that Coulomb's law may be applied to E so as to define it as

$$E = -\nabla v \qquad (2.14)$$

where ∇ = field point gradient operator = $\dfrac{\partial}{\partial x}\hat{x} + \dfrac{\partial}{\partial y}\hat{y} + \dfrac{\partial}{\partial z}\hat{z}$ and $\hat{x}$, $\hat{y}$, and $\hat{z}$ are the spatial coordinates of E

v = scalar potential field function or potential, with units of volts

Substitution of $-\nabla v$ for E in Equation 2.13 yields Equation 2.15:

$$\begin{aligned}
\oint_{l_t} E' \cdot dl &= \int_2^1 -\nabla v \cdot dl \\
&= -v \int_2^1 \nabla \cdot dl \\
&= -v \big|_2^1 \\
&= -(v_1 - v_2) \\
&= v_2 - v_1 \\
&= -v_{\text{loop}} \qquad (2.15)
\end{aligned}$$

Now consider the right-hand side of Faraday's law (Equation 2.8):

$$\frac{-d}{dt}\int_s B \cdot ds \qquad (2.16)$$

The integral indicates the following calculations:

1. At each point within the surface formed by the closed contour calculate the product $B \cdot ds$.
2. Add (integrate) the products.

With the units of webers per square meter and meters squared on B and ds, respectively, the result of the integration is magnetic flux measured in webers.

2.5 APPLICATIONS OF MAXWELL'S EQUATIONS TO MAGNETIC CIRCUITS

$$\int_s B \cdot ds = \phi \quad \text{Wb} \quad (2.17)$$

Therefore, Faraday's law as applied to the loop in Figure 2.6 appears as

$$-v_{\text{loop}} = \frac{-d\phi}{dt} \qquad v_{\text{loop}} = \frac{d\phi}{dt} \quad (2.18)$$

Now, if it is assumed that a flux ϕ passes through all N loops in the coil, the voltage developed across the coil is

$$\begin{aligned} v_{\text{coil}} &= N v_{\text{loop}} \\ &= N \frac{d\phi}{dt} \\ &= \frac{dN\phi}{dt} \\ &= \frac{d\lambda}{dt} \end{aligned} \quad (2.19)$$

where $\lambda = N\phi$ = flux linkage of the coil with units of weber turns (Wbt).

Clearly, in order to calculate v_{coil}, the value of ϕ through the coil loops must be known. Examining Equations 2.6 to 2.10 shows that this information is most likely available from Ampere's law. Magnetic flux appears in the form of H on the left-hand side of Ampere's law and current in the form of J appears as the source of H on the right-hand side.

Ampere's law as shown in Equation 2.6 indicates that a contour of interest must be chosen in the magnetic device. In order to clarify where this contour should be, three assumptions are made about the device.

1. All of the magnetic field flux developed inside the coil remains in the core material of the circuit except when it crosses the air gap. No flux leaks out around the coil turns or strays out of the core material.
2. The air gap g is so small relative to the width W and depth D that the magnetic field flux crosses the air gap in a straight-line path without fringing out to dimensions more than W and D.
3. The permeability μ of the core material is nearly infinite.

The first assumption influences the selection of the closed contour for Ampere's law. Since all of the flux developed by the coil remains in the iron core and air gap, the most convenient and useful contour to select is through the iron core, as shown in Figure 2.9.

Note that the contour has been oriented in a clockwise direction, as indicated by the vector labeled dl in the bottom leg. Also note that this direction is the same as the direction of the magnetic field developed

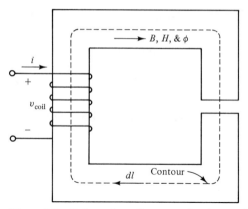

Figure 2.9 Orientation of contour for Ampere's law.

by the coil as indicated by B, H, and ϕ. The magnetic field direction is found by using the right-hand rule on the coil.

The fact that dl was chosen in the same direction as ϕ does not mean that it must always be chosen that way. The contour direction could have been oriented in the opposite direction and the calculations for coil magnetic flux would be the same. However, some additional algebraic complexity occurs in the form of minus signs that appear in subsequent equation development. So as a matter of convenience in the calculations, the contour orientation was selected to match the magnetic field direction.

Now consider the left-hand side of Ampere's law. As stated earlier, it indicates that once a contour has been selected, the product $H \cdot dl$ should be calculated at every point along the path, and then the products should be summed. If the iron core is a homogeneous material, the integration can be broken down into two integrals:

$$\oint_l H \cdot dl = \int H_{iron} \cdot dl_{iron} + \int H_{gap} \cdot dl_{gap}$$
$$= H_{iron} l_{iron} + H_{gap} g \qquad (2.20)$$

where H_{iron} = magnetic field intensity along the iron core path. This quantity is a uniform value everywhere in the iron at each instant in time

l_{iron} = mean core path length

H_{gap} = magnetic field intensity in the air gap. This quantity is a uniform value everywhere in the gap at each instant in time

g = gap width

Now consider the right-hand side of Ampere's law. This part of the equation says to sum (integrate) the current piercing the surface formed by the chosen contour. Examination of Figure 2.9 shows that the only current piercing the surface is in the windings of the coil. Ampere's law calculates the current as $J \cdot ds$ where J is a current density vector and ds

2.5 APPLICATIONS OF MAXWELL'S EQUATIONS TO MAGNETIC CIRCUITS

is an incremental part of the surface. Assuming the turns of wire have a small cross section compared to the entire contour surface, each turn yields a dot product of

$$J \cdot ds = i \quad \text{A} \tag{2.21}$$

Note that special attention must be given to the algebraic sign on i. This sign is determined by the orientation of J with respect to ds.

Vector J has the same direction as current in the turns when the turns pierce the surface formed by the contour. In Figure 2.9 the current in each turn is piercing the surface into the page. Vector ds has a direction determined by the right-hand rule. Using this rule on the contour vector yields ds pointing into the page. Since J and ds have the same direction, their dot product yields a positive value of i amperes.

The total number of products to sum on the right-hand side of Ampere's law is N, the number of turns in the coil. Thus, combining Equations 2.20 and 2.21 yields

$$H_{iron} l_{iron} + H_{gap} g = Ni \tag{2.22}$$

Some simplification of this equation can be achieved by use of the third assumption about the magnetic device, that is, $\mu_{iron} \to \infty$. If $H = B/\mu$ is substituted in Equation 2.22, the result is

$$\frac{B_{iron}}{\mu_{iron}} l_{iron} + \frac{B_{gap}}{\mu_{gap}} g = Ni \tag{2.23}$$

Since $\mu_{iron} \to \infty$ as compared to μ_{gap} (or μ_0), the first term on the left-hand side of Equation 2.23 is small compared to the second term. Thus, Equation 2.23 can be rewritten

$$\frac{B_{gap}}{\mu_{gap}} g = Ni \quad \text{or} \quad \frac{B_{gap}}{\mu_0} g = Ni \tag{2.24}$$

Examination of Equation 2.24 shows that unfortunately information on magnetic flux is available only for the air gap and not for the coil wrapped around the left core leg. The derivation of Faraday's law requires that ϕ for the coil be known. Thus, some kind of relationship between B_{iron} (the magnetic field flowing through the coil) and B_{gap} must be found that can be applied to Equation 2.24. The remaining equation not used in Maxwell's set is the magnetic source law. Application of this equation to the magnetic device is the next step to find the relationship between B_{iron} and B_{gap}.

Equation 2.4 shows that the magnetic source law requires that a closed surface must be chosen, and products of $B \cdot ds$ calculated and summed. Since the relationship between B_{gap} and B_{iron} is desired, the obvious placement of the closed surface should be at an interface point between B_{iron} and B_{gap}. For the device shown in Figure 2.7, this interface point is the air gap. Figure 2.10 shows the placement of the closed surface. Note the surface is a box that encloses all of the area of width W and

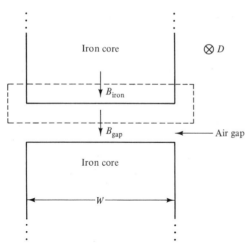

Figure 2.10 Placement of closed surface.

depth D defined by the iron core. It also extends into the iron and into the air gap. Magnetic flux density vectors B_{iron} and B_{gap} are shown entering and leaving the box, respectively.

The magnetic source law applied to the chosen closed surface can be written

$$\oint_s B \cdot ds = \int B_{iron}\, ds_1 + \int B_{gap} \cdot ds_2$$
$$= -|B_{iron}|W \cdot D + |B_{gap}|W \cdot D = 0 \qquad (2.25)$$

where ds_1 = small surface element on the top surface of the box
       ds_2 = small surface element on the bottom surface of the box

No other parts of the surface of the box have magnetic flux piercing them, because the second assumption about the magnetic device states that the flux crossing the air gap does not fringe out at the sides.

Note that the first term of Equation 2.25 has a minus sign. This sign occurs because of the dot product $B_{iron} \cdot ds_1$. Vector B_{iron} is oriented in a direction into the box. Vector ds_1 is oriented in a direction out of the box. This orientation of ds_1 occurs because surface vectors of closed surfaces have been chosen to be oriented pointing out from the surface. Therefore, B_{iron} and ds_1 are pointing in opposite directions, and their dot product yields the minus sign.

Further manipulation of Equation 2.25 yields the desired relationship between B_{iron} and B_{gap}:

$$B_{iron} = B_{gap} \qquad (2.26)$$

Substitution of Equation 2.26 into Equation 2.24 yields

$$\frac{B_{iron}}{\mu_0} g = Ni \qquad (2.27)$$

2.5 APPLICATIONS OF MAXWELL'S EQUATIONS TO MAGNETIC CIRCUITS

Since B_{iron} distributes uniformly through the iron, the flux ϕ through the coil is easily calculated as

$$\phi_{\text{coil}} = B_{\text{iron}} W \cdot D \quad \text{or} \quad B_{\text{iron}} = \frac{\phi_{\text{coil}}}{WD} \tag{2.28}$$

Substitution of Equation 2.28 into Equation 2.27 yields

$$\frac{\phi_{\text{coil}}}{\mu_0 WD} g = Ni \qquad \phi_{\text{coil}} = \frac{\mu_0 WD Ni}{g} \quad \text{Wb} \tag{2.29}$$

Using $\lambda = N\phi$ as noted in Equation 2.19, λ_{coil} appears as

$$\lambda_{\text{coil}} = \frac{\mu_0 WDN^2 i}{g} \quad \text{Wbt} \tag{2.30}$$

At this point the analysis of the device is complete. Given a current i, the voltage induced on the coil is

$$\begin{aligned} v_{\text{coil}} &= \frac{d\lambda_{\text{coil}}}{dt} \\ &= \frac{d}{dt}\left(\frac{\mu_0 WDN^2 i}{g}\right) \\ &= \frac{\mu_0 WDN^2}{g}\frac{di}{dt} \\ &= L\frac{di}{dt} \quad \text{V} \end{aligned} \tag{2.31}$$

where $L = \mu_0 WDN^2/g =$ inductance of the magnetic device, H ■■

Equation 2.31 has been derived by direct application of Maxwell's equation to a magnetic device for which three simplifying assumptions were applied. These assumptions are important points to consider because they do introduce an error into the voltage versus current equation. However, in most cases of interest here, the error is not of significant concern. Another point of interest concerning the expression for ϕ_{coil} in Equation 2.29 involves definition of several terms in common use in electromechanics. Equation 2.29 can be written

$$Ni = \frac{g}{\mu_0 WD}\phi_{\text{coil}} \quad \text{or} \quad \mathscr{F} = \mathscr{R}\phi_{\text{coil}} \tag{2.32}$$

where $\mathscr{F} =$ magnetomotive force or MMF $= Ni$, At

$$\mathscr{R} = \text{reluctance} = \frac{g}{\mu_0 WD}, \text{At/Wb}$$

Equation 2.32 can be rewritten to define permeance:

$$\phi_{\text{coil}} = \frac{\mathscr{F}}{\mathscr{R}} = \mathscr{F}\mathscr{P} \tag{2.33}$$

where $\mathscr{P}$ is the permeance:

$$\mathscr{P} = \frac{1}{\mathscr{R}} = \frac{\mu_0 WD}{g} \quad \text{Wb/At}$$

EXAMPLE 2.2

Another example of a magnetic device is shown in Figure 2.11. Here a movable plunger is surrounded by a stationary yoke with a coil of wire wrapped around its center leg. This device might be used to close or open a set of contacts connected to the plunger whenever current is injected into the coil. The analysis of this device involves solving for the flux linkage λ of the coil. As in the single air gap inductor device, the three assumptions concerning the permeability of the iron, leakage flux around the coil, and fringing of the magnetic field at the air gaps are applied.

Application of Faraday's law to this device parallels exactly its use for the single air gap inductor. It begins by considering a single loop of wire in the coil and calculating v_{coil} in the same manner as in Example 2.1.

$$v_{\text{coil}} = \frac{d\lambda}{dt} \quad \text{V} \tag{2.34}$$

where $\lambda = N\phi$ and ϕ is the flux passing through the coil.

The application of Ampere's law first requires selection of a contour of interest. Since flux through the coil is the quantity desired, and all flux remains in the iron core (except when it crosses air gaps), three contours of interest can be chosen. Figure 2.12 shows these contours. The contours are labeled 1, 2, and 3 and have spatial orientations designated by vectors dl_1, dl_2, and dl_3, respectively. Also note that magnetic field intensities H_1, H_2, and H_3 have been labeled for each of the air gaps. The position of

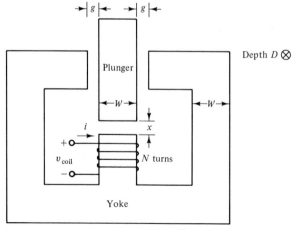

Figure 2.11 Three-air gap solenoid.

2.5 APPLICATIONS OF MAXWELL'S EQUATIONS TO MAGNETIC CIRCUITS

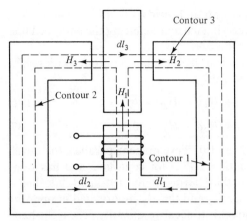

Figure 2.12 Contours of interest in the circuit.

these vectors shows that H_1 is the quantity of concern because the magnetic flux associated with H_1 passes entirely through the coil.

This conclusion can be proven by applying the magnetic source law to air gap x of Figure 2.11, and the iron leg the coil is on.

The first equation written from Ampere's law is for the contour labeled 1.

$$H_{iron1}l_{iron1} + H_1 x + H_2 g = Ni \qquad (2.35)$$

The left-hand side shows three terms. The first term is the product of the magnetic field intensity in the iron part of the contour H_{iron1} multiplied by the mean iron core path length l_{iron1}. The other two terms are the air gap field intensities multiplied by their respective gap lengths. Note that the entire length of the contour is accounted for in the distances l_{iron1}, x, and g, and thus closed-contour integration has been completed around contour 1. Also take special note of the algebraic signs on each of the terms. The left-hand side of Ampere's law requires calculation of the dot product $H \cdot dl$ and careful attention must be given to the orientation of H with respect to dl at every point along the contour of dl. In Figure 2.12 the direction of dl was the same as the assumed directions H_{iron1}, H_1, and H_2 so that their dot products resulted in positive quantities.

The right-hand side of Equation 2.35 is developed from the right-hand side of Ampere's law. Here all of the current piercing the surface formed by contour dl is summed. Again careful attention must be given to the orientation of the current flows with respect to the orientation of the contour surface vector. The right-hand side of Ampere's law requires calculation of the dot product $J \cdot ds$ where J is the current density piercing the surface element ds. Figure 2.12 shows that J is curling across the top of the yoke and back along its right side into the page. The orientation of the surface vector ds is given by applying the right-hand rule to dl_1. Applying this rule to contour 1 shows that ds also points into the page.

Thus, the dot product $J \cdot ds$ for contour 1 yields a positive quantity. The coil has N loops of wire each carrying current i piercing the surface. Thus, the net result of the integration is $+Ni$.

Ampere's law can be applied to the remaining contours with results as shown in Equations 2.36 and 2.37:

$$H_{iron2}l_{iron2} + H_1 x + H_3 g = Ni \tag{2.36}$$

$$H_{iron3}l_{iron3} + H_2 g - H_3 g = 0 \tag{2.37}$$

Equation 2.36 is for contour 2, and its development parallels the development of Equation 2.35. Equation 2.37 is for contour 3 and has some interesting characteristics. The third term on the left-hand side has a negative sign that occurs because H_3 and the dl_3 are oriented in opposite directions. Thus, their dot product is a negative quantity. The right-hand side of Equation 2.37 is zero. This result occurs because the surface formed by dl_3 is pierced by as much current into the page as out of the page. Therefore, the summation of total current piercing the surface in the direction of the surface vector is zero.

At this point the equations are rewritten in terms of magnetic field density quantities using the relationship $B = \mu H$:

$$\frac{B_{iron1}}{\mu_{iron}} l_{iron1} + \frac{B_1}{\mu_0} x + \frac{B_2}{\mu_0} g = Ni \tag{2.38}$$

$$\frac{B_{iron2}}{\mu_{iron}} l_{iron2} + \frac{B_1}{\mu_0} x + \frac{B_3}{\mu_0} g = Ni \tag{2.39}$$

$$\frac{B_{iron3}}{\mu_{iron}} l_{iron3} + \frac{B_2}{\mu_0} g - \frac{B_3}{\mu_0} g = 0 \tag{2.40}$$

Recalling the assumptions that $\mu_{iron} \to \infty$, one sees that the first term in each equation is negligibly small. Since $B = (\phi)(\text{area})$, the three equations appear as

$$\frac{\phi_1}{\mu_0 WD} x + \frac{\phi_2}{\mu_0 WD} g = Ni \tag{2.41}$$

$$\frac{\phi_1}{\mu_0 WD} x + \frac{\phi_3}{\mu_0 WD} g = Ni \tag{2.42}$$

$$\frac{\phi_2}{\mu_0 WD} g - \frac{\phi_3}{\mu_0 WD} g = 0 \tag{2.43}$$

On first glance, the problem of solving for ϕ_1 now appears to be at hand. Equations 2.41 to 2.43 are a set of three equations in three unknowns: ϕ_1, ϕ_2, and ϕ_3. Unfortunately, closer examination reveals that one of the equations is redundant. This is easily seen by subtracting Equation 2.42 from Equation 2.41. The result is Equation 2.43. This revelation parallels the electric circuit rule that the number of independent

2.5 APPLICATIONS OF MAXWELL'S EQUATIONS TO MAGNETIC CIRCUITS

equations is one less than the number of loop equations. Hence, solution of ϕ_1 must involve the selection of a third independent equation. A third independent equation cannot be found by further application of Ampere's law. However, application of the magnetic source law will provide a new independent equation.

The magnetic source law requires, as its first step, the selection of a closed surface of interest within the magnetic device. In this problem the quantities of interest are ϕ_1, ϕ_2, and ϕ_3. Therefore, a surface resulting in an equation including all three quantities is desirable. A good surface for the device under study is shown in Figure 2.13.

The surface selected completely encloses the plunger. The magnetic source law requires surface integration of the dot product of $B \cdot ds$. The three points where B_1, B_2, and B_3 pierce the surface are the only places where a nonzero contribution to the surface integration is made. The equation for the integration appears as

$$\oint_s B \cdot ds = \int_{s1} B_1 \cdot ds_1 + \int_{s2} B_2 \cdot ds_2 + \int_{s3} B_3 \cdot ds_3$$
$$= -B_1 WD + B_2 WD + B_3 WD = 0$$
$$= -\phi_1 + \phi_2 + \phi_3 = 0 \qquad (2.44)$$

Note the first term has a minus sign. This value is a result of the dot product of the integrands $B_1 \cdot ds_1$. The vectors in the product are oriented in opposite directions, and a minus sign results from the dot product. The values of WD in each term come from the areas associated with each air gap and the assumption that the magnetic fields in the air gaps do not fringe outward.

Equation 2.44 and any two equations from Equations 2.41 to 2.43 form a set of three equations in three unknowns, ϕ_1, ϕ_2, and ϕ_3. Solving for ϕ_1 yields

$$\phi_1 = \frac{2WD\mu_0 Ni}{2x + g} \quad \text{Wb} \qquad (2.45a)$$

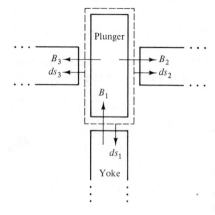

Figure 2.13 Position of closed surface.

At this point reluctance, permeance, and magnetomotive force can be defined for the ϕ_1 equation. With Equation 2.33

$$\mathcal{R} = \frac{2x + g}{2WD\mu_0} \quad \text{At/Wb} \tag{2.45b}$$

$$\mathcal{P} = \frac{2WD\mu_0}{2x + g} \quad \text{Wb/At} \tag{2.45c}$$

$$F = Ni \quad \text{At} \tag{2.45d}$$

Continuing with the solution of λ yields

$$\lambda = N\phi_1 = \frac{2WD\mu_0 N^2 i}{2x + g} \quad \text{Wbt} \tag{2.46}$$

Equation 2.46 completes the analysis of this device as far as flux linkage is concerned. Given a value of i and a position x, the voltage induced in the coil can be found from Faraday's law. For example,

$$v_{\text{coil}} = \frac{d}{dt} \frac{2WD\mu_0 N^2 i}{2x + g}$$

$$= \frac{2WD\mu_0 N^2}{2x + g} \frac{di}{dt} - \frac{4WD\mu_0 N^2 i}{(2x + g)^2} \frac{dx}{dt} \quad \text{V} \quad \blacksquare\blacksquare \tag{2.47}$$

Equation 2.47 illustrates two more points of interest about magnetic devices. The first term on the right-hand side is proportional to di/dt and is called the transformer voltage. The second term on the right-hand side is proportional to dx/dt. This quantity is called the speed voltage because dx/dt is the expression for mechanical speed. The speed term will appear in v_{coil} any time the plunger is in motion. Otherwise, only the transformer voltage will be present for time-varying current.

EXAMPLE 2.3

The same device with two coils is shown in Figure 2.14. The analysis of this device involves solving for the flux linkages λ_1 and λ_2. As in the previous examples, the three assumptions concerning the permeability of the iron, leakage flux around the coils, and fringing of the fields in the air gaps are applied.

Application of Faraday's law to this device is done exactly the same way as in the previous examples. In this example two separate equations, one for each coil, result:

$$v_{\text{coil1}} = \frac{d\lambda_1}{dt} \quad \text{V}$$

$$v_{\text{coil2}} = \frac{d\lambda_2}{dt} \quad \text{V} \tag{2.48}$$

where λ_1 = flux linking coil 1, Wbt
λ_2 = flux linking coil 2, Wbt

2.5 APPLICATIONS OF MAXWELL'S EQUATIONS TO MAGNETIC CIRCUITS

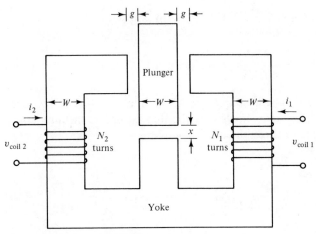

Figure 2.14 Two-coil device.

Application of Ampere's law requires selection of contours of integration. The contours are the same as in the previous example and are shown in Figure 2.12.

Applying Ampere's law to contours dl_1, dl_2, and dl_3, respectively, results in

$$H_{iron1}l_{iron1} + H_1 x + H_2 g = N_1 i_1 \qquad (2.49)$$

$$H_{iron2}l_{iron2} + H_1 x + H_3 g = N_2 i_2 \qquad (2.50)$$

$$H_{iron3}l_{iron3} + H_2 g - H_3 g = N_1 i_1 - N_2 i_2 \qquad (2.51)$$

Substituting H with B/μ, letting $\mu_{iron} \to \infty$, and recalling $B = (\phi)(\text{area})$ yields

$$\frac{\phi_1}{\mu_0 WD} x + \frac{\phi_2}{\mu_0 WD} g = N_1 i_1 \qquad (2.52)$$

$$\frac{\phi_1}{\mu_0 WD} x + \frac{\phi_3}{\mu_0 WD} g = N_2 i_2 \qquad (2.53)$$

$$\frac{\phi_2}{\mu_0 WD} g - \frac{\phi_3}{\mu_0 WD} g = N_1 i_1 - N_2 i_2 \qquad (2.54)$$

Once again, only two of the three equations 2.52 to 2.54 are independent. Therefore, another equation is derived by applying the magnetic source law around the plunger:

$$-\phi_1 + \phi_2 + \phi_3 = 0 \qquad (2.55)$$

Using any two equations from Equations 2.52 to 2.54 and Equation 2.55, one can solve ϕ_1, ϕ_2, and ϕ_3 as

$$\phi_1 = \frac{\mu_0 WD}{(g + 2x)} (N_1 i_1 + N_2 i_2) \qquad (2.56)$$

$$\phi_2 = \frac{\mu_0 WD}{g(g+2x)}[N_1 i_1(x+g) - N_2 i_2 x] \qquad (2.57)$$

$$\phi_3 = \frac{\mu_0 WD}{g(g+2x)}[N_2 i_2(x+g) - N_1 i_1 x] \qquad (2.58)$$

Applying the magnetic source law at the interface of the two upper air gaps and the iron yoke shows that the flux ϕ_2 passes through coil 1 and the flux ϕ_3 passes through coil 2. Therefore,

$$\lambda_1 = N_1 \phi_2$$
$$= \frac{\mu_0 WD}{g(g+2x)}[N_1^2 i_1(x+g) - N_1 N_2 i_2 x] \qquad (2.59)$$

and $\qquad \lambda_2 = N_2 \phi_3$

$$= \frac{\mu_0 WD}{g(g+2x)}[N_2^2 i_2(x+g) - N_1 N_2 i_1 x] \qquad \blacksquare\blacksquare \quad (2.60)$$

Given currents i_1 and i_2 and a position x, voltage V_{coil1} and V_{coil2} can be calculated from Equations 2.48, 2.59, and 2.60.

EXAMPLE 2.4

This example demonstrates the technique of analyzing flux linkage of a rotating device. Figure 2.15 shows a rotating cylinder (rotor) of iron positioned inside a stationary piece (stator) such that a uniform air gap surrounds the rotor. The problem to be solved is an evaluation of the stator

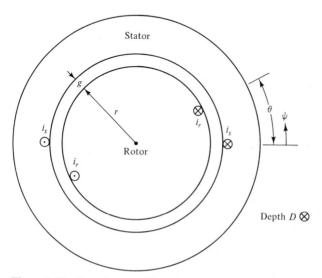

Figure 2.15 Rotating machine.

2.5 APPLICATIONS OF MAXWELL'S EQUATIONS TO MAGNETIC CIRCUITS

coil flux linkage λ_s and the rotor flux linkage λ_r. These two quantities can be used to solve for voltages induced on the coils by Equation 2.61.

$$v_s = \frac{d\lambda_s}{dt} \quad \text{and} \quad v_r = \frac{d\lambda_r}{dt} \tag{2.61}$$

where v_s = stator coil voltage
       v_r = rotor coil voltage

Both coils are assumed to be negligibly thin and wound in slots of negligible depths. The rotor has a radius of r meters and the air gap is g meters wide such that $r \gg g$. Current directions are indicated by crosses and dots inside the coil turns. The wire in the coils travels the depth of the machine, circles around the back, returns on the other side, and closes upon itself across the front of the machine. The three assumptions regarding permeability, leakage flux, and magnetic field fringing are applied to this example.

The previous examples have shown that the air gaps in magnetic devices have a much higher reluctance than the iron because iron permeability is higher than μ_0. Thus, the air gap fields play a major role in analysis with Ampere's law. This example has a device with one continuous air gap, so analysis of its field seems appropriate. Assume that the air gap magnetic field is directed radially outward from the rotor to the stator and is of uniform magnitude along the depth D of the machine. Also assume that the rotor position θ is in the range $0 \leqslant \theta < \pi$. Figure 2.16 shows placement of contours of integration for use in Ampere's law. Contour dl_1 crosses the air gap from the rotor to stator at an angle ψ such that $0 \leqslant \psi < \theta$ and travels through the stator to cross the air gap again at

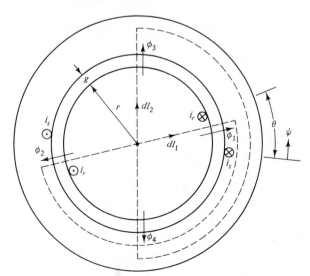

Figure 2.16 Contours of integration for Ampere's law.

$\psi + \pi$. Ampere's law for this contour appears as

$$\oint_l H \cdot dl = \int_s J \cdot ds$$
$$H_{rotor} 2r + H_1 g + H_{stator} l_{stator} - H_2 g = N_s i_s - N_r i_r \qquad (2.62)$$

where H_{rotor} = magnetic field intensity inside the rotor along dl_1
H_{stator} = magnetic field intensity inside the stator
l_{stator} = mean path length of dl_1 inside the stator
N_s = number of turns in stator coil
N_r = number of turns in rotor coil

With $B = \mu H$, Equation 2.62 is rewritten

$$\frac{B_{rotor}}{\mu_{iron}} 2r + \frac{B_1}{\mu_0} g + \frac{B_{stator}}{\mu_{iron}} l_{stator} - \frac{B_2}{\mu_0} g = N_s i_s - N_r i_r \qquad (2.63)$$

By the $\mu_{iron} \to \infty$ assumption, the terms involving B_{rotor} and B_{stator} are negligible, so Equation 2.63 can be reduced to

$$\cancelto{0}{\frac{B_{rotor}}{\mu_{iron}} 2r} + \frac{B_1}{\mu_0} g + \cancelto{0}{\frac{B_{stator}}{\mu_{iron}} l_{stator}} - \frac{B_2}{\mu_0} g = N_s i_s - N_r i_r \qquad (2.64)$$

Note that the position of the dl_1 can be shifted from crossing the air gaps at $\psi = 0$ and $\psi = \pi$ to crossing at $\psi = \theta$ and $\psi = \pi + \theta$ and still leave $N_s i_s - N_r i_r$ as the value on its right-hand side. Figure 2.17 illustrates this idea. This characteristic implies that

$$\begin{aligned} B_1 &= \text{constant} && \text{for } 0 \leq \psi \leq \theta \\ B_2 &= \text{constant} && \text{for } \pi \leq \psi \leq \pi + \theta \end{aligned} \qquad (2.65)$$

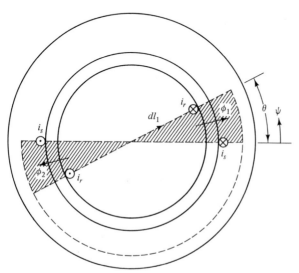

Figure 2.17 Alternate positions of contour dl_1.

2.5 APPLICATIONS OF MAXWELL'S EQUATIONS TO MAGNETIC CIRCUITS

Equation 2.65 can be used to calculate values of ϕ_1 and ϕ_2:

$$\phi_1 = B_1(\text{area}_1)$$
$$= B_1 2\pi r \frac{\theta}{2\pi} D$$
$$= B_1 r\theta D \quad 0 \leq \psi \leq \theta \quad (2.66)$$
$$\phi_2 = B_2 r\theta D \quad \pi \leq \psi \leq \pi + \theta$$

Substituting the results of Equation 2.66 into Equation 2.64 yields

$$\frac{\phi_1 g}{\mu_0 r\theta D} - \frac{\phi_2 g}{\mu_0 r\theta D} = N_s i_s - N_r i_r \quad (2.67)$$

Corresponding development of contour dl_2 in Figure 2.16 gives

$$\frac{\phi_3 g}{\mu_0 r(\pi - \theta)D} - \frac{\phi_4 g}{\mu_0 r(\pi - \theta)D} = N_s i_s + N_r i_r \quad (2.68)$$

Equations 2.67 and 2.68 form a set of two equations in four unknown quantities—ϕ_1, ϕ_2, ϕ_3, and ϕ_4. In order to solve for any one of the unknowns, at least two more independent equations must be found.

Several additional equations can be created by using Ampere's law on four more contours shown in Figure 2.18. These equations are developed in the same manner as Equations 2.67 and 2.68.

for contour dl_3 $$\frac{-\phi_1}{\mu_0 \varphi \theta D} g + \frac{\phi_3}{\mu_0 r(\pi - \theta)D} g = N_r i_r \quad (2.69)$$

for contour dl_4 $$\frac{\phi_1}{\mu_0 r} g - \frac{\phi_4}{\mu_0 r(\pi - \theta)D} g = N_s i_s \quad (2.70)$$

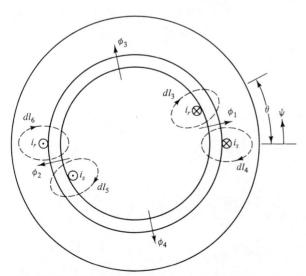

Figure 2.18 Additional contours of integration for Ampere's law.

for contour dl_5 $\quad \dfrac{\phi_4}{\mu_0 r(\pi - \theta)D} g - \dfrac{\phi_2}{\mu_0 r\theta D} g = -N_r i_r \quad$ (2.71)

for contour dl_6 $\quad \dfrac{\phi_2}{\mu_0 r\theta D} g - \dfrac{\phi_3}{\mu_0 r(\pi - \theta)D} g = -N_s i_s \quad$ (2.72)

Equations 2.69 to 2.72 can be reduced to two equations by adding (a) Equation 2.69 to 2.71 and (b) Equation 2.70 to 2.72:

(a) $\quad \dfrac{\phi_1}{\mu_0 r\theta D} g - \dfrac{\phi_2}{\mu_0 r\theta D} g + \dfrac{\phi_3}{\mu_0 r(\pi - \theta)D} g + \dfrac{\phi_4}{\mu_0 r(\pi - \theta)D} g = 0 \quad$ (2.73)

(b) $\quad \dfrac{\phi_1}{\mu_0 r\theta D} g + \dfrac{\phi_2}{\mu_0 r\theta D} g - \dfrac{\phi_3}{\mu_0 r(\pi - \theta)D} g - \dfrac{\phi_4}{\mu_0 r(\pi - \theta)D} g = 0 \quad$ (2.74)

The results of these additions show that Equations 2.73 and 2.74 are redundant. Therefore, repeated application of Ampere's law to the rotating machine yields a set of three equations (Equations 2.67, 2.68, and 2.73 or 2.74) in four unknowns. A fourth independent equation must be found.

The previous examples have shown that the last independent equation involving flux is found by applying the magnetic source law. Application of this law requires selection of a closed surface. The most useful surface for the rotating machine is a cylinder that surrounds the rotor, as shown in Figure 2.19. Since no flux fringes out of the machine at each end, only the radial contribution of flux appears in the surface integral.

$$\oint_s \mathbf{B} \cdot d\mathbf{s} = \int \mathbf{B}_1 \cdot d\mathbf{s}_1 + \int \mathbf{B}_2 \cdot d\mathbf{s}_2 + \int \mathbf{B}_3 \cdot d\mathbf{s}_3 + \int \mathbf{B}_4 \cdot d\mathbf{s}_4 = 0 \quad (2.75)$$
$$0 = B_1 r\theta D + B_2 r\theta D + B_3 r(\pi - \theta)D + B_4 r(\pi - \theta)D$$
$$= \phi_1 + \phi_2 + \phi_3 + \phi_4$$

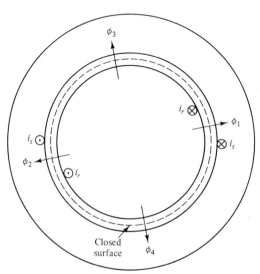

Figure 2.19 Position of surface for magnetic source law.

2.5 APPLICATIONS OF MAXWELL'S EQUATIONS TO MAGNETIC CIRCUITS

Equations 2.67, 2.68, 2.73 or 2.74, and 2.75 form a set of four independent equations in four unknowns. The solution for ϕ_1, ϕ_2, ϕ_3, and ϕ_4 yields

$$\phi_1 = \frac{\mu_0 r \theta D}{2g}(N_s i_s - N_r i_r) \tag{2.76}$$

$$\phi_2 = \frac{-\mu_0 r \theta D}{2g}(N_s i_s - N_r i_r) \tag{2.77}$$

$$\phi_3 = \frac{\mu_0 r (\pi - \theta) D}{2g}(N_s i_s + N_r i_r) \tag{2.78}$$

$$\phi_4 = \frac{-\mu_0 r (\pi - \theta) D}{2g}(N_s i_s + N_r i_r) \tag{2.79}$$

Equations 2.76 and 2.77 show an interesting feature about the relationship between coil-winding distribution and radially directed air gap flux. If the coil windings are symmetrically distributed around the machine, the fluxes at diametrically opposed points in the air gap will be equal in magnitude but opposite in direction. This phenomenon is clearly exhibited in the machine shown in Figure 2.15 because its windings are symmetrical and $\phi_1 = -\phi_2$ and $\phi_3 = -\phi_4$.

The solution for λ_s and λ_r follow easily from Equations 2.76 to 2.79. The total flux passing through the stator coil is simply

$$\phi_s = \phi_1 + \phi_3$$
$$= \frac{\mu_0 r D}{2g}[\pi N_s i_s + (\pi - 2\theta)N_r i_r] \quad \text{for} \quad 0 \leq \theta \leq \pi \tag{2.80}$$

The value of λ_s follows from $\lambda = N\phi$:

$$\lambda_s = N_s \phi_s$$
$$= \frac{\mu_0 r D}{2g}[\pi N_s^2 i_s + (\pi - 2\theta)N_s N_r i_r] \quad \text{for} \quad 0 \leq \theta \leq \pi \tag{2.81}$$

The value of λ_r follows from

$$\lambda_r = N_r \phi_r$$
$$= N_r(\phi_3 + \phi_2)$$
$$= \frac{\mu_0 r D}{2g}[\pi N_r^2 i_r + (\pi - 2\theta)N_s N_r i_s] \quad \text{for} \quad 0 \leq \theta \leq \pi \tag{2.82}$$

For θ in the range of $\pi \leq \theta < 2\pi$, calculation of air gap flux yields

$$\lambda_s = \frac{\mu_0 r D \pi}{2g} N_s^2 i_s + \frac{\mu_0 r D}{2g}(2\theta - 3\pi)N_r N_s i_r \tag{2.83}$$

$$\lambda_r = \frac{\mu_0 r D}{2g}(2\theta - 3\pi)N_r N_s i_s + \frac{\mu_0 r D \pi}{2g} N_r^2 i_r \quad \blacksquare\blacksquare \tag{2.84}$$

2.6 MAGNETIC FIELD ENERGY

Force developed in devices such as relays, solenoids, and so forth is caused by a magnetic field acting on a magnetizable material. Since some kind of motion occurs in these devices, they are using mechanical energy. The principle of conservation of energy states that energy cannot be destroyed or created, only changed in form. When this principle is applied to magnetic devices, the source of the mechanical energy output must be electric energy input, and the mechanism through which the electric energy is transformed to mechanical energy is the magnetic field in the device. Thus, in order to quantify the mechanical force developed by a magnetic device, the energy stored in its magnetic field must be investigated.

The investigation of magnetic field energy will be confined to lossless field devices. This consideration does not impose severe restrictions on its application to a magnetic device because any losses that actually exist in the device can be lumped externally in its input and output circuits. Figure 2.20 shows a block diagram of a magnetic field device. The figure shows the magnetic field device in the center block with electrical excitation of voltage v and current i and with mechanical excitation of force of electrical origin f^e and mechanical displacement x. If the energy stored in the magnetic field is denoted W_m, any change in either electric energy input or mechanical energy output will be accompanied by a corresponding change in W_m. This change has to occur because the magnetic field is transforming electric energy to mechanical energy or vice versa. Equation 2.85 expresses the change in W_m as a differential, dW_m, in terms of the mechanical and electric energy:

$$dW_m = dW_{elec} - dW_{mech} \qquad (2.85)$$

where dW_m = change in stored energy of magnetic field
 dW_{elec} = change in electric energy input
 dW_{mech} = change in mechanical energy output

In order to develop this expression further, the terms on the right-hand side must be expressed in terms of the variables in Figure 2.20. Equation 2.86 shows this development for dW_{elec}:

$$\begin{aligned} dW_{elec} &= vi\,dt \\ &= \frac{d\lambda}{dt} i\,dt \\ &= d\lambda\, i \end{aligned} \qquad (2.86)$$

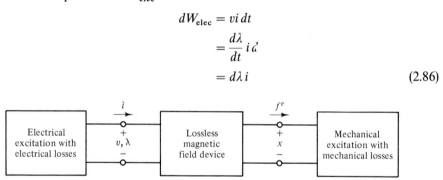

Figure 2.20 Block diagram of magnetic field device.

2.6 MAGNETIC FIELD ENERGY

Here the power input, vi, which is the rate at which energy is consumed or produced, is multiplied by a change in time, dt, to yield the change in electric energy. Appropriate transformation of v using Faraday's law yields the final expression.

Equation 2.87 shows the corresponding development for mechanical energy:

$$dW_{\text{mech}} = f^e u \, dt$$
$$= f^e \frac{dx}{dt} dt$$
$$= f^e \, dx \tag{2.87}$$

where u is the velocity. Here the mechanical power, $f^e u$, is multiplied by a change in time, dt, to yield the change in mechanical energy. Appropriate transformation of u using dx/dt yields the final expression.

Substitution of Equations 2.86 and 2.87 into Equation 2.85 results in Equation 2.88:

$$dW_m = i \, d\lambda - f^e \, dx \tag{2.88}$$

In working with this equation to solve for W_m, attention must be turned to the selection of independent variables. Consideration of the physical characteristics of the device shows that, of the four variables i, λ, f^e, and x, only two may be selected as independent. For example, previous work shows that λ and i are a function of each other. Thus, selecting λ as an independent variable excludes i being independent and vice versa.

Since an incremental change in W_m is shown in Equation 2.88 to be in part a function of incremental changes in λ and x, these variables will be selected as independent. Solution for W_m using this selection requires an integration through the variable space of λ and x from a given starting point to a given finishing point. This operation is a line integration and requires selection of a path of integration from the starting point to the finishing point. However, a property of a lossless system is that its energy state is not dependent upon the path chosen to reach that state. Therefore, integration from an initial energy state to a final energy state should be performed on a path that is mathematically convenient. Figure 2.21 shows an example of an energy calculation.

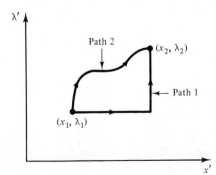

Figure 2.21 Example energy calculation.

Two paths are shown. Using path 1 to calculate the magnetic energy integral is easier because the independent variables x and λ do not change simultaneously. Both paths of integration yield the same change in energy. However, path 2 requires an additional preliminary step to change all of the functions in the integrand to a common variable. Thus, path 1 is more mathematically convenient.

Equation 2.88 presents what might appear to be a significant problem in calculating ΔW_m. It requires an explicit expression for $f^e(\lambda, x)$. In Examples 2.1 to 2.4 only a function $i(x, \lambda)$ is available; that is, $f^e(\lambda, x)$ is unknown. This problem is easily resolved if the magnetic device under study is energized from $(x', \lambda') = (0, 0)$ to $(x', \lambda') = (x, \lambda)$ where (x, λ) is a final operating point of interest. The definition of f^e is that it is a "force of electrical origin," which means that f^e does not exist unless electric excitation has been applied to the device. This fact can be used to determine a path of integration from $(0, 0)$ to (x, λ) over which $f^e = 0$ or $dx = 0$. Then, the ΔW_m calculation will be a function of only λ. The following example illustrates this concept.

EXAMPLE 2.5

Consider the three-air gap solenoid shown in Example 2.2. Calculate the ΔW_m for an initial operating point $(x, \lambda) = (0, 0)$ to a final point (x, λ). The calculation for λ yielded

$$\lambda(x, i) = \frac{2WD\mu_0 N^2 i}{2x + g} \tag{2.89}$$

Solving for $i(x, \lambda)$ yields

$$i(x, \lambda) = \frac{(2x + g)\lambda}{2WD\mu_0 N^2} \tag{2.90}$$

Figure 2.22 shows the independent-variable space (x, λ) and the chosen path of integration.

Note that on the horizontal part of the path $f^e(x', \lambda') = 0$ because no electrical excitation has been applied to the circuit; that is, $\lambda' = 0$. On the vertical part of the path, x' is a constant value of x and $dx = 0$. The

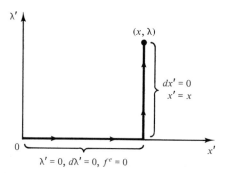

Figure 2.22 Path of integration for three-air gap solenoid.

2.7 FORCE CALCULATION FROM THE EXPRESSION FOR MAGNETIC FIELD ENERGY

calculation for ΔW_m under these conditions is

$$\Delta W_m = \int_{(0,0)}^{(x,\lambda)} i(x',\lambda')\,d\lambda' - f^e(x',\lambda')\,dx'$$

$$= \int_{(0,0)}^{(x,0)} \overset{0}{i(x',\lambda')\,d\lambda'} - f^e(x',\lambda')\,dx' + \int_{(x,0)}^{(x,\lambda)} i(x',\lambda')\,d\lambda' - \overset{0}{f^e(x',\lambda')\,dx'}$$

(integration along horizontal part)   (integration along vertical part)

$$= \int_{(x,0)}^{(x,\lambda)} i(x',\lambda')\,d\lambda'$$

$$= \int_0^\lambda \frac{(2x+g)}{2WD\mu_0 N^2}\,\lambda'\,d\lambda'$$

$$= \frac{(2x+g)(\lambda)^2}{4WD\mu_0 N^2} \qquad \blacksquare\blacksquare \quad (2.91)$$

Equation 2.91 shows that careful selection of the path of integration eliminates the portions of the integration requiring $f^e(x,\lambda)$ and results in an integral evaluation of $i(x,\lambda)$ only. Another way of looking at this calculation is to think of the magnetic device as having been put together mechanically before electric excitation, after which the device is held in its final mechanical position while its electric excitation is applied. The horizontal part of the path is the mechanical construction. The vertical part of the path is the subsequent electric excitation.

2.7 FORCE CALCULATION FROM THE EXPRESSION FOR MAGNETIC FIELD ENERGY

Since force acting on the pieces of a magnetic device is of interest, a method of calculating forces for the devices is desirable. A positive force of electrical origin acts in a direction that will increase the displacement of movable pieces. This rule holds because of the minus sign on $f^e\,dx$ in Equation 2.88. The minus sign indicates that energy is being removed from the magnetic field and transferred into the mechanical circuit. A transfer of energy into the mechanical circuit causes a force to act over a displacement. Hence f^e causes displacement(s) to occur in an increasing direction.

To calculate f^e, one uses the differential of W_m written in terms of the independent variables λ and x.

$$dW_m = \frac{\partial W_m}{\partial \lambda}\,d\lambda + \frac{\partial W_m}{\partial x}\,dx \qquad (2.92)$$

Combining this equation with Equation 2.88 yields

$$0 = \left(i - \frac{\partial W_m}{\partial \lambda}\right)d\lambda - \left(f^e + \frac{\partial W_m}{\partial x}\right)dx \qquad (2.93)$$

Since λ and x are independent variables, their values are unrestricted. Thus, Equation 2.93 can be satisfied only if the coefficients of $d\lambda$ and dx are both 0. This condition yields the desired relationship between energy and force:

$$f^e = -\frac{\partial W_m}{\partial x} \quad (2.94)$$

and it also yields

$$i = \frac{\partial W_m}{\partial \lambda} \quad (2.95)$$

EXAMPLE 2.6

Calculate the f^e for the magnetic device shown in Example 2.2. The first step in the calculation is to show the direction of f^e in the device. Figure 2.23 illustrates its position. Force f^e must always be placed in a position that will increase the displacement of moving pieces. Therefore, it is shown to be acting in a direction to increase x.

The value of W_m for this circuit was calculated in Example 2.5 as

$$W_m = \frac{(2x + g)\lambda^2}{4WD\mu_0 N^2} \quad (2.96)$$

Applying Equation 2.94 to this result yields

$$f^e = \frac{-\lambda^2}{2WD\mu_0 N^2} \quad \blacksquare \quad (2.97)$$

Note that Equation 2.97 shows f^e to have a negative value. This means that f^e actually acts in a direction opposite to the direction shown in Figure

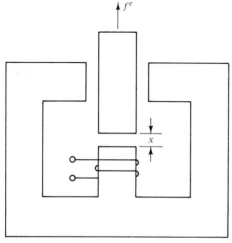

Figure 2.23 Direction of f^e.

2.8 FORCE AND COENERGY RELATIONSHIPS

2.23. That is, it acts to close the gap x when current is applied to the coil. This action is a characteristic of magnetic devices. Magnetic devices always tend to reduce their reluctance.

2.8 FORCE AND COENERGY RELATIONSHIPS

One problem that exists in calculating f^e from W_m is that W_m requires an expression for i in terms of λ. However, the usual method of evaluating a magnetic device yields λ in terms of i. This condition means that an initial step of converting λ in terms of i into i in terms of λ must be taken before evaluating W_m. If more than one source of electric excitation exists in the device, this conversion can be somewhat involved. An alternative method of evaluating f^e exists that avoids this initial step.

The alternative method uses a new variable, called coenergy and designated as W'_m. Coenergy is derived by switching the independent variables from λ, x to i, x by using differentials on Equation 2.88. Recall that $d(i\lambda)$ is

$$d(i\lambda) = i\,d\lambda + \lambda\,di \tag{2.98}$$

Solving for $i\,d\lambda$,

$$i\,d\lambda = d(i\lambda) - \lambda\,di \tag{2.99}$$

and substituting for it in Equation 2.88 yields

$$dW_m = d(i\lambda) - \lambda\,di - f^e\,dx \tag{2.100}$$

$$-d(W_m - i\lambda) = \lambda\,di + f^e\,dx \tag{2.101}$$

From Equation 2.101 coenergy is defined as

$$W'_m = i\lambda - W_m \tag{2.102}$$

and

$$dW'_m = \lambda\,di + f^e\,dx \tag{2.103}$$

Equation 2.103 represents the coenergy differential and can be used to calculate the change in coenergy from a line integral in the variable space i and x.

The calculation of f^e from Equation 2.103 follows that for f^e calculation derived from W_m. The differential of W'_m is written

$$dW'_m = \frac{\partial W'_m}{\partial i}di + \frac{\partial W'_m}{\partial x}dx \tag{2.104}$$

Combining this equation with Equation 2.103 yields

$$0 = \left(\frac{\partial W'_m}{\partial i} - \lambda\right)di + \left(\frac{\partial W'_m}{\partial x} - f^e\right)dx \tag{2.105}$$

Since i and x are independent variables, their coefficients must be 0 in order to satisfy Equation 2.105. The relationships between W'_m and i and W'_m and

f^e are easily derived:

$$f^e = \frac{\partial W'_m}{\partial x} \tag{2.106}$$

and

$$\lambda = \frac{\partial W'_m}{\partial i} \tag{2.107}$$

The expression for dW'_m changes somewhat when more than one electric or mechanical port is present. A contribution to dW'_m comes from each port. In general, a device with m electric ports and n mechanical ports has a coenergy differential, as shown in Equation 2.108:

$$dW'_m = \sum_{j=1}^{m} i_j d\lambda_j + \sum_{k=1}^{n} f_k^e dx_k \tag{2.108}$$

and

$$f_k^e = \frac{\partial W'_m}{\partial x_k} \tag{2.109}$$

EXAMPLE 2.7

Calculate f^e using the coenergy approach for the magnetic device of Example 2.3.

Recall that the flux linkages were solved as

$$\lambda_1 = \frac{\mu_0 WD}{g(g+2x)}[N_1^2 i_1(x+g) - N_1 N_2 i_2 x] \tag{2.110}$$

$$\lambda_2 = \frac{\mu_0 WD}{g(g+2x)}[N_2^2 i_2(x+g) - N_1 N_2 i_1 x] \tag{2.111}$$

Using the path of integration shown in Figure 2.24, solve for W'_m.

$$\begin{aligned}
W'_m &= \int_{(0,0,0)}^{(x,i_1,i_2)} \lambda_1 \, di'_1 + \lambda_2 \, di'_2 + f^e \, dx' \\
&= \int_{0|i'_2=0}^{i_1} \lambda_1 \, di'_1 + \int_{0|i'_1=i_1}^{i_2} \lambda_2 \, di'_2 \\
&= \frac{\mu_0 WD}{2g(g+2x)}[N_1^2 i_1^2(x+g) + N_2^2 i_2^2(x+g) - 2N_1 N_2 i_1 i_2 x]
\end{aligned} \tag{2.112}$$

Now, using Equation 2.109 to solve for f^e and letting $i'_1 = i_1$ and $i'_2 = i_2$, there results

$$f^e = \frac{-\mu_0 WD}{2(g+2x)^2}(N_1 i_1 + N_2 i_2)^2 \tag{2.113}$$

where the direction of f^e is in the same direction as positive x. ■■

A rotating electromechanical device develops a torque or torques of electrical origin, τ^e. Deriving an expression for τ^e in terms of W'_m begins by replacing translational displacement x with rotational displacement θ in the

2.8 FORCE AND COENERGY RELATIONSHIPS

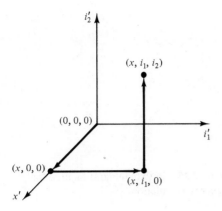

Figure 2.24 Path of integration.

expression for dW_{mech}. The arc length of rotational motion is defined as

$$x = r\theta \tag{2.114}$$

where r is the moment arm and θ is the angle of rotation. Substituting for x in Equation 2.87 results in

$$dW_{mech} = f^e\, dx = f^e\, d(r\theta) \tag{2.115}$$

Since r is constant

$$dW_{mech} = f^e r\, d\theta \tag{2.116}$$

By definition, torque equals force times moment arm. Therefore,

$$dW_{mech} = f^e\, dx = \tau^e\, d\theta \tag{2.117}$$

Substituting into Equation 2.103 for $f^e\, dx$ yields

$$dW'_m = \lambda\, di + \tau^e\, d\theta \tag{2.118}$$

For multiple-port systems of m electrical ports and n mechanical ports

$$dW'_m = \sum_{j=1}^{m} \lambda_j\, di_j + \sum_{k=1}^{n} \tau^e_k\, d\theta_k \tag{2.119}$$

and

$$\tau^e_k = \frac{\partial W'_m}{\partial \theta_k} \tag{2.120}$$

EXAMPLE 2.8

Calculate τ^e using the coenergy approach for the magnetic circuit for Example 2.4 for $0 \leqslant \theta \leqslant \pi$. Recall that the flux linkages for this problem were solved as

$$\lambda_s = \frac{\mu_0 Dr\pi N_s^2 i_s}{2g} + \frac{\mu_0 Dr(\pi - 2\theta)N_s N_r i_r}{2g} \tag{2.121}$$

$$\lambda_r = \frac{\mu_0 Dr(\pi - 2\theta)N_s N_r i_s}{2g} + \frac{\mu_0 Dr\pi N_r^2 i_r}{2g} \tag{2.122}$$

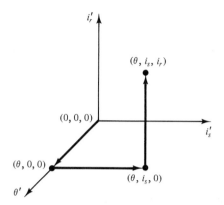

Figure 2.25 Path of integration used for coenergy calculation.

Using the path of integration shown in Figure 2.25, solve for W'_m.

$$W'_m = \int_{(0,0,0)}^{(\theta,i_s,i_r)} \lambda_s \, di'_s + \lambda_r \, di'_r + \tau^e \, d\theta'$$

$$= \int_0^{i_s} \lambda_s(\theta, i'_s, 0) \, di'_s + \int_0^{i_r} \lambda_r(\theta, i_s, i'_r) \, di'_r$$

$$= \frac{\mu_0 Dr}{2g} \left[\frac{\pi N_s^2 (i_s)^2}{2} + (\pi - 2\theta) N_s N_r i_r i_s + \frac{\pi N_r^2 (i_r)^2}{2} \right] \quad (2.123)$$

The calculation of τ^e uses Equation 2.120:

$$\tau^e = \frac{-\mu_0 Dr N_s N_r i_s i_r}{g} \quad \blacksquare\blacksquare \quad (2.124)$$

2.9 SUMMARY

In this chapter the characteristics of magnetic devices and electromechanical devices have been examined. Methods have been developed to analyze the conversion of electric energy to and from mechanical energy. Equations have been derived that directly relate mechanical forces and electrical variables. The analytical tools that were presented and used in this chapter are the basis for understanding the operation of most electric power devices and, therefore, are important concepts for all electric power engineers. These basic concepts will be used in later chapters as new elements of the power system are introduced.

2.10 PROBLEMS

Assume the following magnetic field conditions in the following problems:

(a) Permeability of the iron is nearly infinite; that is, $\mu_{iron} \to \infty$.
(b) Air gap distances are so small relative to their cross-sectional areas that magnetic flux crossing them does not fringe out.

2.10 PROBLEMS

(c) The magnetic flux remains in the iron except when it crosses the air gaps.

2.1. Given the magnetic field device shown in Figure 2.26, calculate the flux linkage λ of the coil. The movable plunger is constrained to move only up or down.

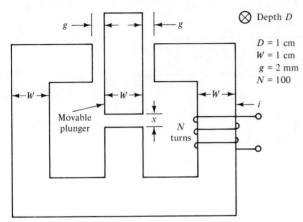

Figure 2.26 Magnetic device.

2.2. Given the magnetic field device shown in Figure 2.27, calculate the flux linkage λ of the coil. The movable block is constrained to move only up, down, left, or right.

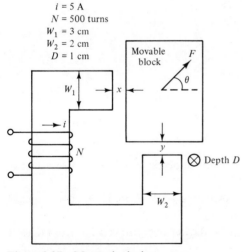

Figure 2.27 Magnetic device.

2.3. Given the magnetic field device shown in Figure 2.28, calculate the flux linkage of both coils, λ_1 and λ_2. The movable block is constrained to move only left or right.

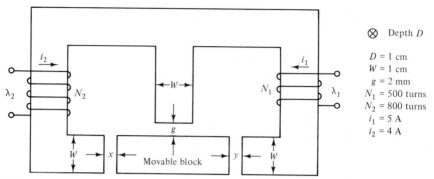

Figure 2.28 Magnetic device.

2.4. Calculate the flux linkages λ_A, λ_B, and λ_R for the magnetic field device shown in Figure 2.29 for θ in the range of $0 \leqslant \theta \leqslant \pi/2$.

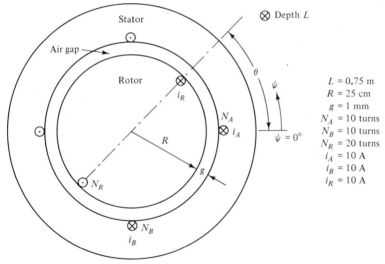

Figure 2.29 Magnetic device.

2.5. Calculate the magnetic field energy for the device of Problem 2.1. Use values of $x = 1$ mm and $i = 10$ A.

2.6. Calculate the magnetic field energy for the device of Problem 2.2. Use values of $x = 1$ mm and $y = 2$ mm.

2.7. Calculate the magnetic field coenergy for the device of Problem 2.1. Use values of $x = 1$ mm and $i = 10$ A.

2.8. Calculate the magnetic field coenergy for the device of Problem 2.2. Use values of $x = 1$ mm and $y = 2$ mm.

2.9. Calculate the magnetic field coenergy for the device of Problem 2.3. Use values of $x = 1$ mm and $y = 3$ mm.

2.10 PROBLEMS

2.10. Calculate the magnetic field coenergy for the device of Problem 2.4. Use a value of $\theta = 35°$.

2.11. Calculate the force of electrical origin acting on the plunger of the device in Problem 2.1. Use the operating conditions of Problem 2.5. Specify both direction and magnitude for the force.

2.12. Calculate the forces of electrical origin acting on the movable block of the device in Problem 2.2. Use the operating conditions of Problem 2.6. Specify both direction and magnitude of the forces.

2.13. Calculate the forces of electrical origin acting on the movable block of the device in Problem 2.3. Use the operating conditions of Problem 2.9. Specify both direction and magnitude of the forces.

2.14. Calculate the torque of electrical origin acting on the rotor of the device in Problem 2.4. Use the operating conditions of Problem 2.10. Specify both direction and magnitude of the torque.

Chapter 3

Synchronous Generators

3.1 INTRODUCTION

Power systems have evolved from a collection of isolated systems into vast interconnected networks of generating stations and transmission lines that span whole continents. Many companies own and operate either wholly or in partnership the components making up the network. Each company has the responsibility to provide safe, adequate, and satisfactory electrical service to all customers who seek it within their respective service territories. These responsibilities begin at the generating stations where energy sources in various forms are converted into electric energy. Because almost all of the generating stations are interconnected through a network of transmission lines, uniformity in the operation of generating units is absolutely essential in order to provide satisfactory electrical service. This uniform operation includes strict control over voltage magnitudes and frequency through a wide range of customer load conditions.

Synchronous machines exhibit properties that make them suitable for the generation of electric energy. For this reason nearly all generating stations use them in some combination with a source of mechanical rotating energy. Figure 3.1 shows a functional diagram of the basic elements used in a generating station.

The source of mechanical rotating energy in Figure 3.1 is a steam-powered turbine. The turbine receives steam from a fossil-fueled boiler or perhaps a nuclear reactor boiler. The steam flow into the turbine is controlled by a valve at which steam pressures of 2000 psi and temperatures of 1000 °F are common. The turbine shown in Figure 3.1 is labeled as a "turbine set" in order to emphasize that in large generating stations, several separate turbines may be in the

3.1 INTRODUCTION

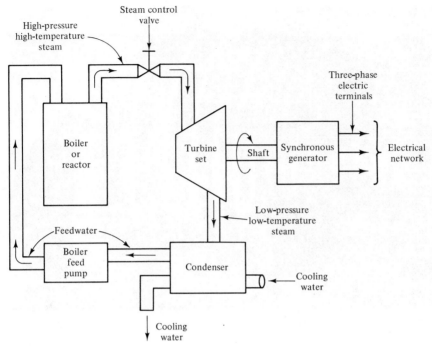

Figure 3.1 Functional diagram of a generating station.

steam flow path. Typical combinations of turbines include high-pressure, intermediate-pressure, and low-pressure turbines that may all be mounted on and driving a single shaft or mounted on and driving, in some combination, multiple shafts. Figure 3.2 shows a turbine with its casing removed. The blades connected to the shaft may be several feet long. Figure 3.3 shows a low-pressure turbine with its casing covering the shaft.

Steam leaves the turbine set at lower temperature and pressure and travels through a condenser where it is cooled until it condenses into water. The water leaving the condenser is pumped back to the boiler by a series of pumps. These pumps are a critical part of generating station operation. Two or more of them may be operated in parallel with an additional pump held in reserve to replace any of the others that break down.

The feedwater that enters the boiler can be heated by the flame from fossil fuel (e.g., coal) being burned or from heat released by controlled nuclear reaction in a nuclear reactor. In the case of a fossil-fueled boiler, the water passes through several heating stages usually culminating in a superheater section, by which time it has been vaporized into high-pressure and high-temperature steam. It is then fed back into the turbine set through the control valve. Figure 3.4 shows a detailed schematic diagram of a fossil-fueled cyclone boiler. Note that the vertically aligned water tube jackets are nearly 200 ft tall.

Returning to Figure 3.1, the synchronous generator is shown connected to the same shaft as the turbine set. The generator converts the spinning energy

Figure 3.2 Steam turbine with casing removed.

Figure 3.3 Steam turbine with casing intact.

3.1 INTRODUCTION

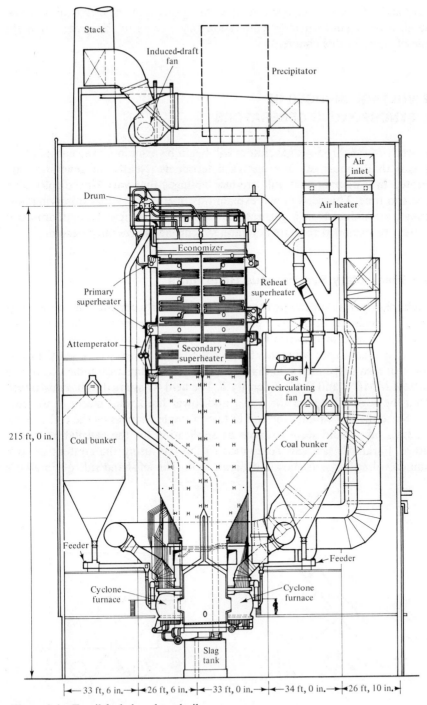

Figure 3.4 Fossil-fueled cyclone boiler.

3.2 VOLTAGE INDUCED IN SYNCHRONOUS GENERATORS

The study of synchronous generators will begin by examining the nature of the voltages that appear at their electrical terminals. Nearly all generators are operated in a manner that will produce voltages that vary sinusoidally with respect to time. This section will explain this operation by first observing how voltages are induced on a generator of elementary design. That information will then be extended to more common synchronous generator designs.

3.2.1 Elementary Synchronous Generators

Generators operate on the principle of an electrically conducting coil moving through a magnetic field. These concepts will be discussed with the help of the elementary generator shown in Figure 3.5. Figures 3.5a and b show top and front views of a square coil, which is able to rotate around an axis called the rotor shaft. It it also suspended in a magnetic field with a direction indicated by the resultant MMF R and the flux density B. The field is assumed to be constrained within the region between the upper and lower horizontal pieces of the coil.

Since the flux linking the coil in this example changes as the coil rotates, Faraday's law would seem to have an application for the calculation of coil voltage. Faraday's law can be applied to the coil using the contour dl with assumed voltage v_{coil}, as shown in Figure 3.5b. The left-hand side of Faraday's

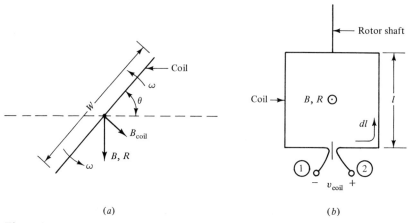

Figure 3.5 View of elementary machine from (a) top and (b) front.

3.2 VOLTAGE INDUCED IN SYNCHRONOUS GENERATORS

law is calculated as in Chapter 2:

$$\oint_l E \cdot dl = \int_1^2 E \cdot dl + \int_2^1 E \cdot dl$$
$$= \int_1^2 E \cdot dl \quad \text{(because the coil is a perfect conductor)}$$
$$= \int_1^2 -\nabla v \cdot dl \quad \text{(because of static field conditions at 1 and 2)}$$
$$= -(v_2 - v_1)$$
$$= -v_{coil} \tag{3.1}$$

The right-hand side of Faraday's law becomes

$$-\frac{d}{dt}\int_s B \cdot ds = -\frac{d\lambda_{coil}}{dt} \tag{3.2}$$

where λ_{coil} is the flux linkage of the coil. Equating the results of Equations 3.1 and 3.2 yields

$$v_{coil} = \frac{d\lambda_{coil}}{dt} \tag{3.3}$$

For the coil shown in Figure 3.5, the flux linkage is

$$\lambda_{coil} = N\phi_{coil} = \phi_{coil} \tag{3.4}$$

because the number of turns N, in the coil is 1. The flux passing through the coil, ϕ_{coil}, is found from

$$\phi_{coil} = B_{coil}(\text{area}) = B_{coil}(Wl) \tag{3.5}$$

The flux density, B_{coil}, is the component of B in Figure 3.5 that is normal to the plane of the coil. Trigonometric construction shows that B_{coil} is calculated as

$$B_{coil} = B \cos \theta \tag{3.6}$$

and with the coil rotated at speed ω such that $\theta = \omega t$, B_{coil} appears as

$$B_{coil} = B \cos \omega t \tag{3.7}$$

Gathering the results of Equations 3.4 to 3.7 into Equation 3.3 yields v_{coil}:

$$v_{coil} = \frac{d\lambda_{coil}}{dt}$$
$$= \frac{d}{dt} WlB \cos \omega t$$
$$= -WlB\omega \sin \omega t \tag{3.8}$$

EXAMPLE 3.1

The coil shown in Figure 3.6 is rotating in a magnetic field density of 2.0 Wb/m^2. Its rotational velocity is 2 rad/s. Plot the voltage induced on the coil terminals for two complete revolutions of the coil.

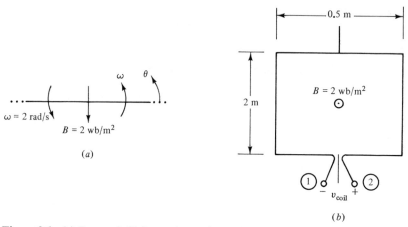

Figure 3.6 (a) Top and (b) front views of rotating coil.

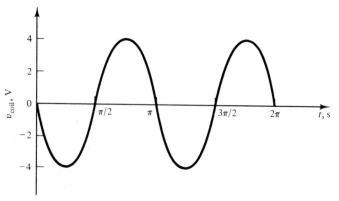

Figure 3.7 Plot of v_{coil}.

On the assumption that the position of the coil shown in Figure 3.6 occurs at $t = 0$, the voltage induced on the coil is given by Equation 3.8.

$$v_{coil} = -(0.5)(2.0)(2.0)(2.0) \sin 2t \quad \text{V}$$
$$= -4 \sin 2t \quad \text{V}$$

The coil completes one revolution in

$$t = [2\pi \text{ (rad/rev)}][\tfrac{1}{2} \text{ (s/rad)}] = \pi \text{ s/rev}$$

Plotting voltage for two revolutions results in Figure 3.7. ■■

3.2.2 Voltage Induced in a Synchronous Generator

Equation 3.8 shows that the induced voltage of the elementary machine is sinusoidal. This kind of voltage is desirable for power system operation because it can be raised and lowered in magnitude by devices called transformers. Chap-

3.2 VOLTAGE INDUCED IN SYNCHRONOUS GENERATORS

ter 4 will show that this kind of control is advantageous because it helps to reduce power losses in the system. Thus, sinusoidal voltages are desirable for synchronous generator operation, and the design of the generator shown in Figure 3.5 might appear to be suitable. However, the movement of the coil means that some kind of mechanical circuit linkage to the stationary world is needed to tap the voltage off the rotating shaft. Slip rings fixed on the rotor shaft contacted by sliding metal brushes fixed to the stationary part of the machine is a method that can tap the voltage. Unfortunately, this technique has some undesirable qualities:

1. Slip rings would be another part of the machine that could break down and would require periodic maintenance.
2. Synchronous generators are often built with heavy-duty-current requirements that slip rings might not be able to withstand owing to arcing and pitting of the brush–slip-ring contact point.

For these reasons synchronous generators have a design that appears as if the elementary machine has been turned inside out. That is, the applied field is established in part by a rotating coil, and the machine's source voltage is induced on stationary coils surrounding the rotating coil. While slip rings are still required for the applied field current, this current is usually much smaller than load current.

Figure 3.8 shows a possible design for a synchronous generator. The rotating piece or rotor is suspended inside a stationary piece called the stator. It has coils or windings embedded in slots in its outer surface, and the wires in the slots run down the length l of the rotor and wrap around each end. Note that the coils on the rotor are distributed around its surface. This distribution is designed to promote a sinusoidal variation of the rotor magnetic field within the air gap. This kind of variation helps to induce sinusoidal voltages in the windings shown on the stationary piece of Figure 3.8.

The stationary piece or stator has three sets of windings or phases labeled a, b, and c. The windings are shown to be centered 120° apart around the inner surface of the stator. Each winding provides one phase of the voltage required for three-phase loads (see Appendix B for a discussion of three-phase voltage and currents). The three currents that flow into the loads are labeled i_a, i_b, and i_c. Figure 3.9 shows an example of a synchronous machine stator. In this device the three phases are distributed over more than one slot.

The voltages that appear on the stator windings are found in accordance with Faraday's law. For example, the voltage induced on winding or phase a appears as

$$v_a = \frac{d\lambda_a}{dt} \tag{3.9}$$

If a synchronous generator is not connected to an electric load, i_a, i_b, and i_c are all zero. Therefore, the flux linking the stator coils is a result of the rotor current only. The flux created by the rotor current will flow outward across

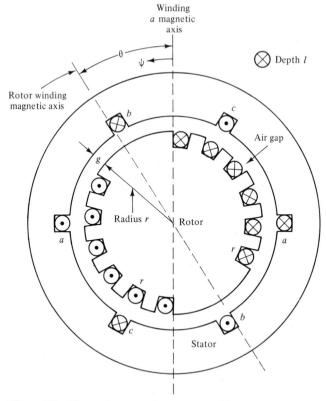

Figure 3.8 Three-phase synchronous machine.

half of the air gap of the machine in Figure 3.8, through the stator iron, and back inward across the other half of the air gap. The plane of the rotor coil's magnetic axis will separate the inward and outward halves of the air gap.

As noted earlier, the rotor windings are distributed around the rotor surface such that the flux will be distributed sinusoidally around the air gap. Therefore, the flux linkages of each stator coil will be a sinusoidal function of rotor position.

The flux linkages of the stator coils are

$$\lambda_a = L_{sr} \cos \theta \, i_r \tag{3.10}$$

$$\lambda_b = L_{sr} \cos(\theta - 120°) i_r \tag{3.11}$$

$$\lambda_c = L_{sr} \cos(\theta - 240°) i_r \tag{3.12}$$

$$\lambda_r = L_r i_r \tag{3.13}$$

where L_{sr} is the mutual inductance between the rotor and stator coils and L_r is the self-inductance of the rotor coil.

If i_r is a constant dc value, I_r, and the rotor is turning at ω_m radians per second with an initial displacement of α, application of Faraday's law to λ_a

3.2 VOLTAGE INDUCED IN SYNCHRONOUS GENERATORS

Figure 3.9 Three-phase stator.

will yield v_a:

$$v_a = \frac{d}{dt}(L_{sr}I_r \cos\theta)$$
$$= \frac{d}{dt}[L_{sr}I_r \cos(\omega_m t + \alpha)]$$
$$= -\omega_m L_{sr}I_r \sin(\omega_m t + \alpha)$$
$$= \omega_m L_{sr}I_r \cos(\omega_m t + \alpha + 90°) \quad (3.14)$$

which is the desired sinusoidally varying voltage on phase a. Voltages on the other phases are

$$v_b = \omega_m L_{sr}I_r \cos(\omega_m t + \alpha - 30°)$$
$$v_c = \omega_m L_{sr}I_r \cos(\omega_m t + \alpha - 150°)$$

Inspection of v_a, v_b, and v_c shows that they are all cosine functions with frequencies which are dependent upon the mechanical speed of the rotor. Therefore, if the flux density is sinusoidal, the synchronous machine design presented is capable of generating sinusoidal voltage with a dc rotor current. The voltages calculated here are the voltages induced on each phase winding under no-load conditions.

EXAMPLE 3.2

Calculate λ_a and v_a at $t = 2$ ms for the conditions of $L_{sr} = 0.02$ H, $I_R = 10$ A, $\theta = \omega_m t$, $\omega_m = 377$ rad/s, and $\alpha = 0°$.

For λ_a, using Equation 3.10 yields

$$\lambda_a = (0.02)(10)\cos[(377)(0.002)]$$
$$= 0.1458 \text{ Wbt}$$

For v_a, using Equation 3.14 yields

$$v_a = (377)(0.02)(10)\cos[(377)(0.002) + \pi/2]$$
$$= -51.62 \text{ V}$$ ∎∎

EXAMPLE 3.3

Repeat Example 3.2 for the same conditions, but with $\omega_m = 0$. For λ_a, using Equation 3.10 yields

$$\lambda_a = (0.02)(10)\cos[(0)(0.002)]$$
$$= 0.2 \text{ Wbt}$$

For v_a, using Equation 3.14 yields

$$v_a = (0)(0.02)(10)\cos[(0)(0.002) + \pi/2]$$
$$= 0 \text{ V}$$

This example illustrates the meaning of Faraday's law as applied to synchronous generators. Even though the flux linking winding a is not zero, the voltage induced on winding a is zero because its flux linkage is constant with respect to time. This condition occurs because the coefficient of time in the cosine function, ω_m, is zero. Therefore, unless the rotor is spinning, that is, $\omega_m \neq 0$, open circuit stator voltage will not be induced on any of the stator windings no matter how large the λ's are. ∎∎

EXAMPLE 3.4

Repeat Example 3.2 for the same conditions, but with $I_r = 20$ A. For λ_a, using Equation 3.10 yields

$$\lambda_a = (0.02)(20)\cos[(377)(0.002)]$$
$$= 0.2916 \text{ Wbt}$$

For v_a, using Equation 3.14 yields

$$v_a = -103.24 \text{ V}$$

3.3 TORQUE IN SYNCHRONOUS GENERATORS

This example illustrates the relationship between field winding current, i_r, and stator voltages. Changes in voltages for open-circuited stator windings are directly proportional to changes in field current, and in actual practice field current is one of the control elements of synchronous generators. Further analysis of synchronous generators made later in this chapter will show that the reactive power flow into and out of a generator also has a strong relationship with the level of field current. ■■

3.3 TORQUE IN SYNCHRONOUS GENERATORS

Torque developed by a synchronous generator is to a large extent due to current flow in its windings. This torque of electrical origin, τ^e, is directed in opposition to the mechanical torque applied by the prime mover (e.g., steam turbine). In steady-state operation these torques balance each other, and power flows smoothly from mechanical form to electrical form.

Analysis of τ^e is of interest because it provides insight into the electromechanical nature of synchronous generators. This section will derive an expression for τ^e using the coenergy technique described in Chapter 2. This technique defines τ^e as

$$\tau^e = +\frac{\partial W'_m}{\partial \theta} \tag{3.15}$$

where W'_m = magnetic field coenergy
θ = angular displacement of the synchronous generator rotating piece (rotor)

The magnetic field coenergy W'_m is calculated by line integration of

$$dW'_m = \lambda_a\, di_a + \lambda_b\, di_b + \lambda_c\, di_c + \lambda_r\, di_r + \tau^e\, d\theta \tag{3.16}$$

where $\lambda_a, \lambda_b, \lambda_c, \lambda_r$ = flux linkages of windings a, b, c, r, respectively
i_a, i_b, i_c, i_r = currents of windings a, b, c, t, respectively
τ^e = torque of electrical origin
θ = displacement angle of the rotor

Equation 3.16 shows that expressions for flux linkages of all the windings in a generator must be known. Furthermore, these expressions must be for the case of currents flowing in all four windings simultaneously. Thus, the expressions shown in Equations 3.10 to 3.13 are not applicable here because they were derived by assuming that i_a, i_b, and i_c were all zero.

The following sections will derive the expression for τ^e by first developing equations for $\lambda_a, \lambda_b, \lambda_c$, and λ_r. These quantities will be used to calculate an expression for W'_m in terms of θ. Calculation of τ^e will follow immediately by applying Equation 3.15. The physical significance of τ^e relative to the rotation of the rotor can then be determined.

3.3.1 Flux Linkages in a Synchronous Generator

The structure of the synchronous generator considered here is the relatively simple machine shown in Figure 3.10. The windings on the stator are situated around the stator such that their magnetic axes are displaced from each other by $\pm 120°$. All of these windings have the same number of turns, N_s, and the rotor winding has N_r turns.

Consider calculation of λ_a. Recall from Chapter 2 that flux linkage is defined as

$$\lambda = N\phi \tag{3.17}$$

and for winding a

$$\lambda_a = N_s \phi \tag{3.18}$$

where ϕ is the flux passing through winding a. Equation 3.18 can be expanded by replacing ϕ with an expression for the flux passing through the air gap from the right side to the left side of coil a and from the front to the back of the generator. The expression is a surface integral of air gap flux density B and appears

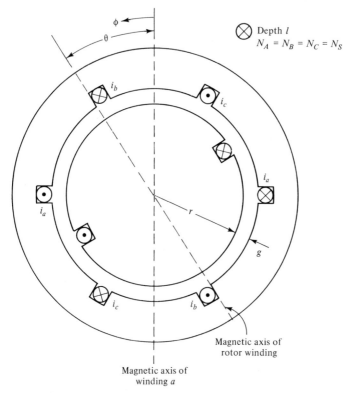

Figure 3.10 Structure of synchronous generator.

3.3 TORQUE IN SYNCHRONOUS GENERATORS

as follows:

$$\lambda_a = N_s \phi$$
$$= N_s \int_{-90°}^{90°} B \cdot ds$$
$$= N_s \int_{-0.5\pi}^{+0.5\pi} \frac{\mu_0}{2g} R\, ds$$
$$= N_s \int_{-0.5\pi}^{+0.5\pi} \frac{\mu_0}{2g} (F_s + F_r)\, ds \quad (3.19)$$

where R = resultant MMF due to both stator and rotor currents
$F_s = F_{sa} + F_{sb} + F_{sc}$ = MMF due to the stator currents
F_{sa}, F_{sb}, F_{sc} = MMFs due to stator currents i_a, i_b, and i_c
F_r = MMF due to the rotor current

The development of Equation 3.19 shows a need for expressions of F_s and F_r in terms of the generator currents. Chapter 2 showed how to calculate these quantities by using Ampere's law on a closed contour oriented as shown in Figure 3.11. This figure shows the contour orientation for calculating F_{sa} and results in

$$H_1 g + H_2 g = N_s i_a \quad 0° \leq \psi < 90° \text{ and } 270° \leq \psi < 360° \quad (3.20)$$
$$H_1 g + H_2 g = -N_s i_a \quad 90° \leq \psi < 270° \quad (3.21)$$

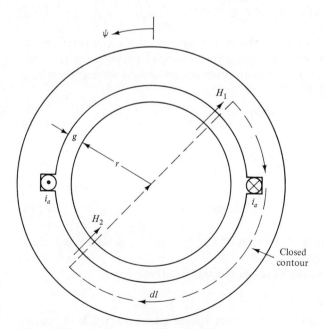

Figure 3.11 Application of Ampere's law.

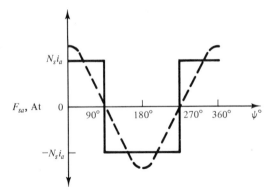

Figure 3.12 F_{sa} plotted with respect to ψ.

The MMF, F_{sa}, is given on the right-hand side of Equations 3.20 and 3.21. Plotting F_{sa} with respect to ψ yields Figure 3.12. Superimposed on the plot of F_{sa} is a cosine wave that can be used as an approximation of the expression for F_{sa}. Actual synchronous generator winding design and construction is nearly always done in such a way as to make the actual MMF distribution more nearly sinusoidal around the air gap. This kind of distribution promotes better synchronous generator operating characteristics, as was noted in Section 3.2.

Using the cosine wave approximation, F_{sa} appears as

$$F_{sa} = \frac{4N_s i_a}{\pi} \cos \psi \tag{3.22}$$

The $4/\pi$ term in the magnitude of the cosine function is the coefficient of the fundamental component of the Fourier series representation of a square wave. The same analysis and approximation provide expressions for the remaining windings.

$$F_{sb} = \frac{4N_s i_b}{\pi} \cos(\psi - 120°) \tag{3.23}$$

$$F_{sc} = \frac{4N_s i_c}{\pi} \cos(\psi - 240°) \tag{3.24}$$

$$F_r = \frac{4N_r i_r}{\pi} \cos(\psi - \theta) \tag{3.25}$$

An interesting feature of the sum of F_{sa}, F_{sb}, and F_{sc}, which was defined as F_s, is that it rotates around the air gap if i_a, i_b, and i_c are balanced three-phase currents. The following example illustrates this feature.

EXAMPLE 3.5

A three-phase synchronous generator is constructed such that the radius of the rotor is $r = 0.5$ m, the length of the rotor is $l = 2.0$ m, the width of the air gap is $g = 0.005$ m, and the number of turns on each stator and

3.3 TORQUE IN SYNCHRONOUS GENERATORS

rotor windings are $N_s = 4$ and $N_r = 200$. It is operated with the following excitation conditions:

$$i_a = 100 \cos(377t) \quad \text{A}$$
$$i_b = 100 \cos(377t - 0.67\pi) \quad \text{A}$$
$$i_c = 100 \cos(377t - 1.33\pi) \quad \text{A}$$
$$i_r = 8 \text{ A}$$
$$\theta = 377t + 0.77\pi \quad \text{rad}$$

Plot F_s for $t = 0$, 2.08, and 6.25 ms.

Substituting the expressions for the current into Equations 3.22 to 3.24 and adding yields F_s.

$$F_s = \frac{(4)(4)(100)}{\pi} [\cos(377t)\cos(\psi) + \cos(377t - 0.67\pi)\cos(\psi - 0.67\pi)$$
$$+ \cos(377t - 1.33\pi)\cos(\psi - 1.33\pi)] \quad \text{At}$$

Using the trigonometric identity $\cos(\alpha)\cos(\beta) = \frac{1}{2}\cos(\alpha - \beta) + \frac{1}{2}\cos(\alpha + \beta)$, the expression for F_s reduces to

$$F_s = \frac{(4)(4)(100)}{2\pi}[3\cos(377t - \psi) + \cos(377t + \psi) + \cos(377t + \psi - 1.33\pi)$$
$$+ \cos(377t + \psi - 2.66\pi)] \quad \text{At}$$

The last three terms within the brackets sum to zero for all t since they are phase shifted by 120° with respect to each other. F_s can be written in final form

$$F_s = 764 \cos(377t - \psi) \quad \text{At}$$

Substituting the required values for t and plotting with respect to ψ yields the graphs shown in Figure 3.13.

Examination of Figure 3.13 shows that as t increases, the peak positive point of F_s shifts to the right to increasingly larger values of ψ. Furthermore, F_s continues this shifting or rotation such that the length of time required for its peak positive point to make one complete revolution is exactly equal to the period of the currents i_a, i_b, and i_c. That is, the MMF wave, F_s, rotates around the air gap at the synchronous speed corresponding to the stator currents frequency. ■■

EXAMPLE 3.6

Plot the resultant MMF, $R = F_s + F_r$, for the conditions of Example 3.5.

The expression for F_r is given in Equation 3.25. Substituting the rotor operating conditions into this equation yields

$$F_r = 2037 \cos(377t + 0.77\pi - \psi) \quad \text{At}$$

Adding F_s from Example 3.5 to F_r results in

$$R = 764 \cos(377t - \psi) + 2037 \cos(377t + 0.77\pi - \psi) \quad \text{At}$$

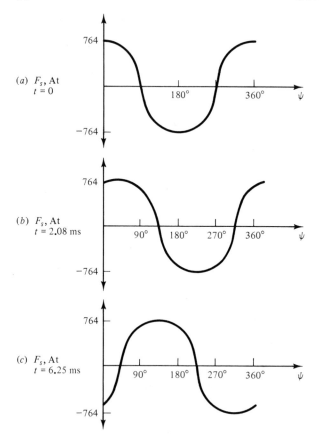

(a) F_s, At $t = 0$

(b) F_s, At $t = 2.08$ ms

(c) F_s, At $t = 6.25$ ms

Figure 3.13 Plots of F_s versus ψ for $t = 0$, 2.08, and 6.25 ms.

which reduces to

$$R = 1548 \cos(377t + 0.66\pi - \psi) \qquad \text{At}$$

Substituting the required values for t and plotting with respect to ψ yields the graphs shown in Figure 3.14. Examination of Figure 3.14 shows that as t increases, the peak positive point of R shifts to the right to increasingly larger values of ψ. And, as in Example 3.5, its period of shifting or rotation is equal to the period of the stator currents i_a, i_b, and i_c.

The fact that the resultant MMF, R, rotates about the rotor axis is an important concept. As was demonstrated in Section 3.2.1 with the elementary synchronous generator, a rotating magnetic field is essential for inducing sinusoidal voltage on generator coils. The rotation of R meets this requirement for the synchronous generator of Figure 3.10. Note that the rotation is caused by the combined effects of the rotation of F_r and F_s. Rotation of F_r is, of course, caused by the physical rotation of the rotor shaft. On the other hand the rotation of F_s is caused by the spacings

3.3 TORQUE IN SYNCHRONOUS GENERATORS

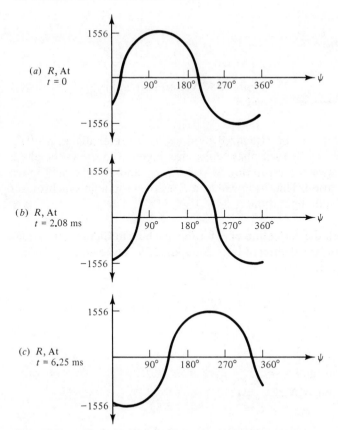

Figure 3.14 Plots of R versus ψ for $t = 0$, 2.08, and 6.25 ms.

of coils a, b, and c at equal intervals around the stator and energizing them with currents that are equally separated in phase. ■■

EXAMPLE 3.7

Determine the phase relationships of F_s, F_r, and R. From Examples 3.5 and 3.6, the equations for F_s, F_r, and R are as follows:

$$F_s = 764 \cos(377t - \psi)$$
$$F_r = 2037 \cos(377t - \psi + 0.77\pi)$$
$$R = 1548 \cos(377t - \psi + 0.66\pi)$$

If F_s is chosen as reference, then F_r leads F_s by

$$0.77\pi - 0 = 0.77\pi \quad \text{rad} \quad \text{or} \quad 138.6°$$

and R leads F_s by

$$0.66\pi - 0 = 0.66\pi \quad \text{rad} \quad \text{or} \quad 118.8°$$

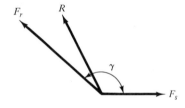

Figure 3.15 Vector diagram for F_s, F_r, and R.

This relationship can be visualized by drawing a vector diagram of F_s, F_r, and R displaced by their phase difference. Figure 3.15 shows the phase relationship between the rotating MMFs F_s, F_r, and R for synchronous generator operation. That is, F_r will lead F_s as they rotate in synchronism during steady-state operation. ■■

Continuing with the derivation of λ_a for the machine in Figure 3.10, Equations 3.22 to 3.25 can be substituted into Equation 3.19. The evaluation of the integral yields the expression for λ_a.

$$\lambda_a = \frac{2\mu_0 N_s^2 rl}{\pi g}(2i_a - i_b - i_c) + \frac{4\mu_0 N_s N_r rl i_r \cos\theta}{\pi g}$$
$$= L_s(2i_a - i_b - i_c) + L_{sr} i_r \cos\theta \qquad (3.26)$$

where
$$L_s = \frac{2\mu_0 N_s^2 rl}{\pi g}$$

$$L_{sr} = \frac{4\mu_0 N_s N_r rl}{\pi g}$$

The remaining flux linkages are solved the same way:

$$\lambda_b = L_s(-i_a + 2i_b - i_c) + L_{sr} i_r \cos(\theta - 120°) \qquad (3.27)$$
$$\lambda_c = L_s(-i_a - i_b + 2i_c) + L_{sr} i_r \cos(\theta - 240°) \qquad (3.28)$$
$$\lambda_r = L_{sr} i_a \cos(\theta) + L_{sr} i_b \cos(\theta - 120°) + L_{sr} i_c \cos(\theta - 240°) + L_r i_r \qquad (3.29)$$

where
$$L_r = \frac{4\mu_0 N_r^2 rl}{\pi g}$$

3.3.2 Calculation of Magnetic Field Coenergy and τ^e

The solution for coenergy W'_m can be performed by line integration of Equation 3.16 over the variable space i'_a, i'_b, i'_c, i'_r, θ' as indicated in Equation 3.30:

$$W'_m = \int_{(0,0,0,0)}^{(i_a,i_b,i_c,i_r)} \lambda_a \, di'_a + \lambda_b \, di'_b + \lambda_c \, di'_c + \lambda_r \, di'_r + \int_0^\theta \tau^e \, d\theta' \qquad (3.30)$$

As was discussed in Chapter 2, the line of integration can be chosen such that the torque integration is always 0. In addition, the currents are independent variables so a line of integration can be chosen where only one of the currents vary at a time. For instance, the first term i'_a can be integrated from 0 to i_a

3.3 TORQUE IN SYNCHRONOUS GENERATORS

while all other currents are held at 0. The second term is then integrated over i'_b from 0 to i_b with i'_a equal to i_a and the other currents at 0. This procedure easily integrates Equation 3.30 and yields

$$W'_m = \frac{L_s}{2} i_a^2 - \frac{L_s}{2} i_a i_b + \frac{L_s}{2} i_b^2 - \frac{L_s}{2} i_a i_c - \frac{L_s}{2} i_b i_c$$

$$+ \frac{L_s}{2} i_c^2 + L_{sr} i_a i_r \cos(\theta) + L_{sr} i_b i_r \cos(\theta - 120°)$$

$$+ L_{sr} i_c i_r \cos(\theta - 240°) + \frac{L_r}{2} i_r^2 \qquad (3.31)$$

Solving for τ^e using Equation 3.15 yields

$$\tau^e = +\frac{\partial W'_m}{\partial \theta}$$

$$= -L_{sr} i_a i_r \sin(\theta) - L_{sr} i_b i_r \sin(\theta - 120°) - L_{sr} i_c i_r \sin(\theta - 240°) \qquad (3.32)$$

In order to have steady-state conversion of mechanical energy to electrical energy, the mechanical torque supplied by the prime mover must be balanced by the torque of electrical origin acting in the opposite direction. However, with normal excitation and operating conditions of

$$i_a = I_{max} \cos \omega_s t$$
$$i_b = I_{max} \cos(\omega_s t - 120°)$$
$$i_c = I_{max} \cos(\omega_s t - 240°)$$
$$i_r = I_r$$
$$\theta = \omega_m t + \gamma$$

Since γ is the angle between F_s and F_r, the expressions for τ^e in Equation 3.32 will only have a nonzero average value when ω_m equals ω_s. [The proof of this statement is left as an exercise for the student. *Hint:* $\sin X \cos Y = \frac{1}{2}\sin(X + Y) + \frac{1}{2}\sin(X - Y)$.] Therefore, a synchronous generator must operate at a mechanical frequency which equals the electrical frequency of the currents in the stator. Under these conditions the expression for τ^e is

$$\tau^e = \frac{-3L_{sr} I_{max} I_r}{2} \sin \gamma \qquad (3.33)$$

EXAMPLE 3.8

Calculate the mechanical torque applied to the rotor of the machine in Example 3.5.

For steady-state operation the applied mechanical torque is found from

$$\tau_m = -\tau^e$$

$$= \frac{3L_{sr} I_{max} I_r}{2} \sin \gamma$$

Equation 3.26 provides the value of L_{sr} as

$$L_{sr} = \frac{4\mu_0 N_s N_r r l}{\pi g}$$

$$= \frac{(4)(4\pi \times 10^{-7})(4)(200)(0.5)(2.0)}{(\pi)(0.005)}$$

$$= 0.256 \text{ H}$$

Calculation of mechanical torque results in

$$\tau_m = \frac{(3)(0.256)(100)(8)}{2} \sin(0.77\pi)$$

$$= 203.2 \text{ N} \cdot \text{m} \qquad \blacksquare\blacksquare$$

EXAMPLE 3.9

Calculate the input power from the shaft of the machine in Example 3.5. Power is related to torque by

$$P = \tau_m \omega_m = \tau_m \omega_s$$

For the conditions of Example 3.8

$$P = (203.2)(377)$$
$$= 76.6 \text{ kW} \qquad \blacksquare\blacksquare$$

3.3.3 Physical Interpretation of τ^e and γ

The significance of the minus sign on the right-hand side of Equation 3.33 and the physical interpretation of angle γ are explained by first reconsidering the position of the rotor MMF as shown in Figure 3.16. Instantaneous position of the rotor MMF has been defined as $\omega_m t + \gamma$, where γ is the angular position of the rotor MMF with respect to the rotating stator MMF. Angle γ is measured from F_s to F_r and is positive if measured in the same direction as the rotation direction of F_s and F_r. This calculation is shown in Equation 3.34:

$$\text{Angle from } F_s \text{ to } F_r = \text{angle of } F_r - \text{angle of } F_s$$
$$= \omega_m t + \gamma - \omega_s t = \omega_s t + \gamma - \omega_s t$$
$$= \gamma \qquad (3.34)$$

In generator action τ^e must be a negative quantity for transfer of mechanical energy from the rotating shaft into the machine. This requirement means that γ must be greater than zero so that evaluation of Equation 3.33 results in a negative value of τ^e.

Figure 3.17 shows that for generator action γ is greater than zero when F_s lags F_r. Note that τ^e is oriented such that it will increase the displacement angle θ. This assumed direction is required by the expression for coenergy shown in Equation 3.16. However, since τ^e has a negative value for generator action, its actual direction is opposite to the assumed direction shown in Figure

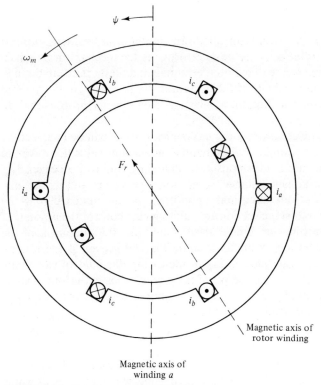

Figure 3.16 Position of rotor MMF.

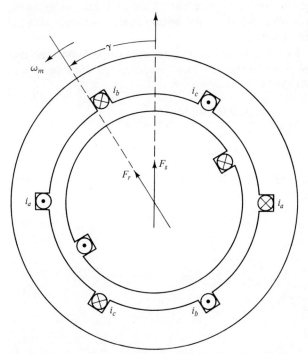

Figure 3.17 Position of F_r with respect to F_s for generator operation.

3.17. Therefore, τ^e acts as a countertorque to the mechanical torque τ_m applied to the rotor by the turbine set. This countertorque is the mechanism by which the electric load connected to the stator terminals manifests itself at the mechanical port (i.e., rotating shaft) of the generator. Its presence can be verified mathematically by noting in Equation 3.33 that τ^e is directly proportional to stator current I_{max}.

Under steady-state conditions, the mechanical torque will be balanced by the load torque τ^e, and the angle γ will remain at a constant value. However, in actual practice, the electric load is constantly changing, and as it does so, I_{max} changes in value to strike new balanced conditions between τ^e and τ_m. As long as this variation is relatively slow and smooth, τ_m can be matched to τ^e by opening and closing the turbine set control valves which control the amount of steam going to the turbine set. Under these conditions the operation of the generator will remain in step with the electric load. However, if the electric load changes rapidly by either decreasing or increasing, the control valve may not be able to match τ_m to τ^e. The result of these conditions is that the turbine set will experience potentially damaging accelerating or decelerating torques as a result of the torque mismatch on the shaft.

EXAMPLE 3.10

Figure 3.18 shows the synchronous generator of Example 3.5 connected to two equal electric loads each drawing current which has a peak value of 50 A. The electric system is operating in steady state when load 1 is suddenly disconnected from the generator. Calculate the accelerating torque experienced by the turbine at the instant load 1 is removed if the rotor current is held at 8 A.

At the instant load 1 is removed, I_{max} reduces to 50 A, and the mechanical torque equivalent to the new load condition is as follows. From Example 3.8, $L_{sr} = 0.256$ H.

$$\tau_{MA} = \frac{(3)(0.256)(50)(8)}{2} \sin(0.77\pi)$$

$$= 101.6 \text{ N·m}$$

which is one-half of the mechanical torque applied to the shaft before load 1 is disconnected, or

$$\tau_{MB} = 203.2 \text{ N·m}$$

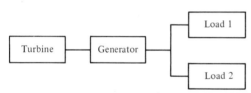

Figure 3.18 Two-load power system.

3.4 MULTIPOLE SYNCHRONOUS GENERATORS

(see Example 3.8). The accelerating torque is the difference between τ_{MB} and τ_{MA}.

$$\tau_{acc} = \tau_{MB} - \tau_{MA}$$
$$= 101.6 \text{ N·m}$$ ■ ■

For the conditions shown, the turbine-generator set will experience, among other effects, an increase in its speed. If the turbine steam control valve cannot close quickly enough to reduce the acceleration, the turbine might experience potentially damaging overspeed vibrations. In most generating stations the turbine speed is monitored for overspeed operation by devices that automatically shut the turbine down to prevent it from being damaged. For the small power system shown in Figure 3.18, this event would result in loss of electric power at load 2, or load 2 will experience a "blackout." For this reason the power system phenomenon associated with sudden load changes is a major area of concern and research in power system engineering.

3.4 MULTIPOLE SYNCHRONOUS GENERATORS

An interesting feature of synchronous generator design is the relationship between speed of rotation of the stator and rotor MMFs, ω_s, and the number of poles per winding. Figures 3.8 and 3.10 show that the synchronous generators being analyzed have one pair of north and south poles per winding, and thus it is said to be a two-pole machine. If the frequency of the electric excitation on the stator is f, the value of ω_s will be

$$\omega_s = \frac{2\pi f}{p/2} \quad \text{rad/s} \tag{3.35}$$

or

$$n_s = \frac{120 f}{p} \quad \text{rev/min (rpm)} \tag{3.36}$$

where f = frequency in hertz
 p = number of poles per winding
 n_s = synchronous speed in revolutions per minute

If the machine in Figure 3.10 has stator currents with a frequency of 60 Hz, ω_s will be 377 rad/s and n_s will be 3600 rpm. Synchronous generators are often built with more than two poles per winding. Correspondingly, a four-pole, 60-Hz synchronous generator would be rotated by a prime mover at 1800 rpm. Figure 3.19 shows the winding placements for a machine with four poles. Notice that the a, b, and c windings each have four poles. The rotor must have the same number of poles as the stator windings. As a result of this placement, the resultant stator MMFs will have two components which are separated by 180 mechanical degrees and the rotor will have two component MMFs also 180° apart.

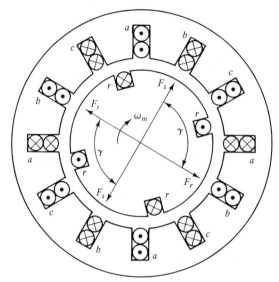

Figure 3.19 Four-pole synchronous machine.

The feature of more than two poles per winding resulting in a lower speed of rotation is used in synchronous generator design for hydroelectric generating stations. These kinds of stations use water-driven turbines that, owing to hydraulic restrictions, operate better at comparatively low speeds. Therefore, synchronous generators used in these applications usually have ten or more poles per winding.

The lower operating speed of the multipole generators will still induce a 60-Hz voltage on the stator windings. For example, the stator windings of a ten-pole generator will experience one complete cycle of the flux wave passing through them for each one-fifth of a mechanical rotation. Since the rotor shaft is spinning at 720 rpm, one-fifth of a mechanical rotation will take 0.0166 s. Therefore, the voltage induced in the stator windings has a period of 0.0166 s, which corresponds to a frequency of 60 Hz.

3.5 EQUIVALENT ELECTRIC CIRCUIT OF A SYNCHRONOUS MACHINE

Analysis of machines in power systems requires that an equivalent electric circuit of the generators be used. The circuit is derived by developing relationships between the terminal voltages of the stator windings and the stator and rotor currents.

The derivation begins by applying Faraday's law to the coil of phase a. Figure 3.20 shows the top view of the coil and the orientation of the contour vector dl and surface vector ds. Chapter 2 provided a derivation of the voltage induced at the terminals of a coil using Faraday's law. Applying that technique

3.5 EQUIVALENT ELECTRIC CIRCUIT OF A SYNCHRONOUS MACHINE

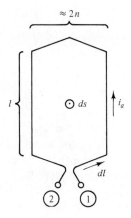

Figure 3.20 Top view of the coil for phase a.

to the coil in Figure 3.20 shows that the left-hand side of Faraday's law reduces to

$$\oint_l E \cdot dl = v_2 - v_1 = v_a \tag{3.37}$$

The right-hand side of Faraday's law reduces to

$$\frac{-d}{dt}\left(\int_s B \cdot ds\right) = \frac{-d}{dt}\lambda_a \tag{3.38}$$

Thus, the voltage induced on phase a appears as

$$v_a = \frac{-d}{dt}\lambda_a \tag{3.39}$$

where λ_a is the same quantity derived in Equation 3.26. The excitation currents and angles cited in the expression for λ_a are assumed to be as follows:

$$i_a = I_{max}\cos(\omega_s t)$$
$$i_b = I_{max}\cos(\omega_s t - 0.66\pi)$$
$$i_c = I_{max}\cos(\omega_s t - 1.33\pi)$$
$$i_r = I_r$$
$$\theta = \omega_m t + \gamma = \omega_s t + \gamma$$

Substitution of these quantities into λ_a and performing the differentiation of Equation 3.39 yields the value for v_a:

$$\begin{aligned}v_a &= 3\omega_s L_s I_{max}\sin(\omega_s t) + \omega_s L_{sr} I_r \sin(\omega_s t + \gamma) \\ &= v_{ar} + v_f\end{aligned} \tag{3.40}$$

where $v_{ar} = 3\omega_s L_s I_{max}\sin(\omega_s t)$ and $v_f = \omega_s L_{sr} I_r \sin(\omega_s t + \gamma)$.

The first term on the right-hand side is called the armature reaction voltage, and the second term is the field excitation voltage. Note that the field excitation voltage v_f is a function of I_r. Thus, v_f is a current-controlled voltage source. A circuit model of phase a using v_{ar} and v_f is shown in Figure 3.21.

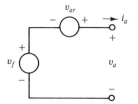

Figure 3.21 Circuit model of phase a using v_f and v_{ar}.

The field excitation voltage is the voltage induced on the stator windings under no-load conditions. The armature reaction voltage represents the effect of MMF created by current flowing in the stator on the resultant MMF. Since i_a is $I_{max}\cos(\omega_s t)$, v_{ar} can be written

$$v_{ar} = -L_m \frac{di_a}{dt} \quad (3.41)$$

where $L_m = 3L_s$ is the magnetizing inductance. Using L_m in the circuit model yields Figure 3.22. The equivalent circuit of Figure 3.22 has been derived by assuming that all of the flux created in the generator links all of the windings. In actual practice this assumption is not valid. Some of the magnetic flux does not link the other windings and hence is referred to as leakage flux. The effect of leakage flux is modeled by placing an additional inductance in the equivalent circuit of the windings. Figure 3.23 shows the additional inductance as L_{la} in the equivalent circuit as well as stator coil series resistance r_a. Usually, the series resistance is small relative to the magnitude of L_m and L_{la}, and for most calculations its effect is negligible. Therefore, the series resistance can be neglected in the equivalent circuit.

The other two stator windings have similar circuits. Because v_a, v_b, and v_c will be balanced three-phase voltages, calculations of voltages and currents on any of the phase circuit models will yield results applicable to the other phases after appropriate $\pm 120°$ phase shifts are applied. Therefore, a more general circuit model is often used as shown in Figure 3.24. Here the phase

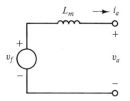

Figure 3.22 Circuit model for phase a using L_m.

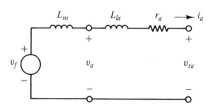

Figure 3.23 Equivalent circuit of phase a with leakage inductance and series resistance.

3.5 EQUIVALENT ELECTRIC CIRCUIT OF A SYNCHRONOUS MACHINE

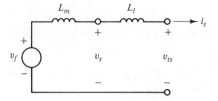

Figure 3.24 General equivalent circuit.

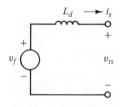

Figure 3.25 Equivalent circuit with synchronous inductance.

subscript have been replaced by the more general subscript s. Inductances L_m and L_l are sometimes combined into a single inductance L_d. This inductance is called synchronous inductance or the direct-axis synchronous inductance. Figure 3.25 shows L_d in the equivalent circuit.

EXAMPLE 3.11

Calculate L_m, L_d, v_f, and v_s for the generator of Example 3.5. Assume that $L_l = 2.65 \times 10^{-4}$ H.

For L_m, use Equation 3.41:

$$L_m = 3L_s$$
$$= 3\frac{2\mu_0 N_s^2 rl}{\pi g}$$
$$= \frac{(3)(2)(4\pi \times 10^{-7})(4^2)(0.5)(2.0)}{(\pi)(0.005)}$$
$$= 0.00768 \text{ H}$$

L_d is calculated from the following expression:

$$L_d = L_m + L_l$$
$$= (76.8 + 2.65) \times 10^{-4}$$
$$= 0.00795 \text{ H}$$

For v_f, use Equation 3.40:

$$v_f = \omega_s L_{sr} I_r \sin(\omega_s t + \gamma)$$
$$= \omega_s \frac{4\mu_0 N_s N_r rl}{\pi g} I_r \sin(\omega_s t + \gamma)$$
$$= \frac{(377)(4)(4\pi \times 10^{-7})(4)(200)(0.5)(2.0)(8)}{(\pi)(0.005)} \sin(377t + 0.77\pi)$$
$$= 772.1 \sin(377t + 0.77\pi)$$

For v_s, use Equations 3.40 and 3.41:

$$v_s = v_f - L_m \frac{di_s}{dt}$$

$$= 772.1 \sin(377t + 0.77\pi) + (377)(0.00768)(100)\sin(377t)$$
$$= 772.1 \sin(377t + 0.77\pi) + 289.5 \sin(377t)$$
$$= 587.0 \sin(377t + 0.66\pi)$$

■■

Up to this point in this chapter all alternating voltages and currents have been represented as peak values multiplied by a sine or cosine function. However, one of the quantities of most interest to an electric power engineer is average power consumed or generated. In order to calculate average power, it is necessary to use the root-mean-squared (rms) values of voltage and current (see Appendix B). Therefore, for convenience, all equivalent circuit voltages and currents shown in Figures 3.21 to 3.25 will be defined as rms phasor quantities. With the exceptions of v_f, v_{ar}, and v_s, all phasors will be denoted by capital V or I with the same subscripts as the instantaneous values they represent. It is a common practice to represent the induced voltages v_f, v_{ar}, v_s in phasor notation as E_f, E_{ar}, E_s. The use of phasor notation for currents and voltages requires that the resistances and inductances be represented as impedances (see Appendix A). Therefore, Figure 3.25 is redrawn using phasor notation in Figure 3.26. The equation which corresponds to Figure 3.26 is

$$E_f = V_{ts} + jX_d I_s \tag{3.42}$$

The standard practice of using rms phasor representation for sinusoidal quantities will be used in the remaining part of this chapter and all following chapters. Unless otherwise specified, all voltages and currents will be given as rms values. For a review of phasors see Appendix A.

Three-phase generators are almost always Y connected. That is, one end of each of the phase coils is connected to a common neutral point. This means that the voltage induced in each coil is a line-to-neutral voltage, and the current flowing in each coil is the current that will flow in each phase line leaving the generator. Therefore, the voltages in the equivalent circuits of Figures 3.21 to 3.26 are line-to-neutral voltages and the currents are line currents. An explanation of line-to-neutral and line-to-line voltages and three-phase calculations is given in Appendix B.

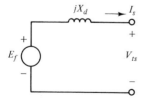

Figure 3.26 Equivalent circuit with phasor notation.

3.5 EQUIVALENT ELECTRIC CIRCUIT OF A SYNCHRONOUS MACHINE

EXAMPLE 3.12

Perform the following calculations for the generator of Example 3.5 (neglect stator leakage reactance):

(a) complex power out for each phase and the power factor
(b) total three-phase complex power output
(c) steady-state mechanical torque acting on the generator

Part A Complex power for each phase is defined in Appendix B as

$$S = V_{l-n} I_l^*$$

where V_{l-n} and I_l are line-to-neutral voltage and line current, respectively. For the generator of Example 3.5, this expression becomes

$$S = E_s I_s^*$$

where E_s and I_s are rms phasor values. Using the results of Example 3.11 yields

$$S = \frac{587.0}{\sqrt{2}} \underline{/118.8°} \left(\frac{100}{\sqrt{2}} \underline{/-90} \right)$$

$$= 29{,}350 \underline{/28.8°} \text{ VA}$$

or

$$S = 25{,}720 + j14{,}140 \text{ VA}$$

The power factor is

$$\text{pf} = \cos(\phi)$$

where ϕ is the angle by which the current lags the voltage. For this case it is calculated as

$$\text{pf} = \cos(28.8°)$$
$$= 0.8763 \text{ lagging}$$

Part B Three-phase power is given as

$$S_T = 3 \times S$$
$$= 3 \times 29{,}350 \underline{/28.8°}$$
$$= 88{,}050 \underline{/28.8°} \text{ VA}$$

or

$$S_T = 77{,}160 + j42{,}420 \text{ VA}$$

Part C Mechanical torque is defined as

$$\tau_m = \frac{P_m}{\omega_m} = \frac{P_T}{\omega_s}$$

where P_T is the three-phase real-power output of the generator as in part B. The mechanical torque is calculated as

$$\tau_m = \frac{77{,}160}{377} = 204.6 \text{ N·m}$$ ■■

3.6 POWER AND TORQUE ANGLE

Additional insight into the operation of synchronous generators can be gained by defining the power or torque angle δ. In terms of the phasor diagram associated with the synchronous generator circuit model, δ is the angle between E_f and V_{ts}, as shown in Figure 3.27. The power output from the rms voltage V_{ts} in the equivalent circuit is related directly to δ. This relationship can be derived as follows:

$$\begin{aligned} P_T &= 3\,\text{Re}\{S\} \\ &= 3\,\text{Re}\{V_{ts}I_s^*\} \\ &= 3\,\text{Re}\left\{V_{ts}\left(\frac{E_f - V_{ts}}{jX_d}\right)^*\right\} \\ &= \frac{3|E_f||V_{ts}|}{X_d}\sin\delta \end{aligned} \qquad (3.43)$$

where δ is positive if V_{ts} lags E_f.

A plot of P_T versus δ for E_f and V_{ts} constant is shown in Figure 3.28. Synchronous generator operation corresponds to positive values of δ to the

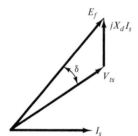

Figure 3.27 Definition of angle δ.

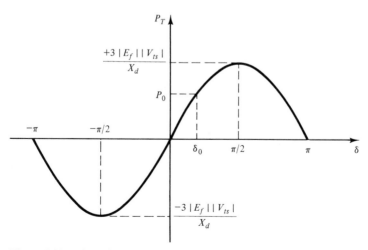

Figure 3.28 Plot of P_T versus δ.

3.6 POWER AND TORQUE ANGLE

right of the P_T axis. Synchronous motor operation corresponds to negative values of δ to the left of the P_T axis.

Consider synchronous generator operation. The P versus δ operating point on the curve will be to the right of the P_T axis. Assume that this point is P_0. If V_{ts}, E_f, and ω_s are held constant, increasing or decreasing demand for power from the generator will be met by increasing or decreasing power angle δ. Suppose the generator experiences an increasing demand for power. The power angle will increase and P_0 will shift to the right on the power curve. When δ reaches 90°, P_0 will be at the peak output power of the generator. Any further increase in demand for power from the generator will result in increasing δ past 90°, and the output power of the generator will fall below the demand. At this point the machine will lose synchronism or pull out of step, probably resulting in the rotor accelerating. This condition causes extreme mechanical stress on the machine and can damage the rotor shaft.

The power at which δ is $\pm 90°$ is called the pullout power and represents a limit on the allowed range of operation for synchronous machines. The magnitude of pullout power varies depending upon the operating conditions imposed on the machine. Different values of V_{ts}, E_f, and ω_s yield different operating conditions.

EXAMPLE 3.13

Draw the three-phase power versus δ curve for the generator of Example 3.5. Identify the operating point (δ_0, P_0) and the generator pullout power.

From the equivalent circuit

$$V_{ts} = E_s - jX_I I_s$$
$$= 415\underline{/118.8°} - j(0.1)(70.7\underline{/+90°})$$
$$= 412\underline{/118°}$$

Using Equation 3.43 and the result of Example 3.11, the P_T versus δ equation appears as

$$P_T = \frac{3|E_f||V_{ts}|}{X_d} \sin \delta$$

$$= \frac{3}{2.995}(546)(412) \sin \delta \quad \text{(recall that } E_f \text{ and } V_{ts} \text{ are rms quantities)}$$

$$= 225{,}330 \sin \delta \quad \text{W}$$

The P_T versus δ plot is shown in Figure 3.29. The value of δ_0 for the conditions of Example 3.5 is

$$\delta_0 = \underline{/E_f} - \underline{/V_{ts}}$$
$$= 138° - 118°$$
$$= 20°$$

P_0 is calculated as

$$P_0 = 225{,}330 \sin(20°)$$
$$= 77{,}067 \text{ W}$$

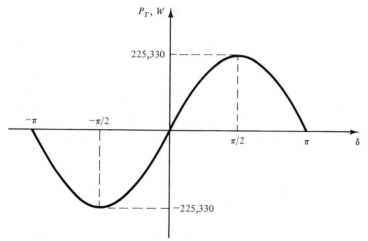

Figure 3.29 P_T versus δ.

which is reasonably close to the result of Example 3.12. The pullout power of the generator is its peak power of 225,330 W. ∎

Equation 3.43 can be used to define the relationship between torque and angle δ. Since $\tau = P/\omega$, this relationship is

$$\tau_m = \frac{P_T}{\omega_s}$$

$$= \frac{3|E_f||V_{ts}|}{\omega_s X_d} \sin \delta \qquad (3.44)$$

3.7 SYNCHRONOUS GENERATORS IN A POWER SYSTEM

In a power system all generators are electrically connected. The system frequency and voltages are determined by the combined action of all the generators. For a power system that has a large number of generators, the frequency and terminal voltage experienced by any single generator are largely independent of any variation in operation by that single generator. Thus, in Equation 3.43, three-phase power output (neglecting stator series resistance) of a synchronous generator is

$$P_T = \frac{3|E_f||V_{ts}|}{X_d} \sin \delta$$

$$= K_1 |E_f| \sin \delta \qquad (3.45)$$

where

$$K_1 = \frac{3|V_{ts}|}{X_d}$$

3.7 SYNCHRONOUS GENERATORS IN A POWER SYSTEM

This equation shows that for real-power control the generator operator must control $|E_f|$ or δ. A similar conclusion can be drawn for three-phase reactive-power control. Three-phase reactive power is expressed

$$Q_T = \frac{3|V_{ts}||E_f|}{X_d} \cos \delta - \frac{3|V_{ts}|^2}{X_d}$$
$$= K_1|E_f| \cos \delta - K_2 \tag{3.46}$$

where
$$K_2 = \frac{3|V_{ts}|^2}{X_d}$$

If δ is changed an incremental amount, Equations 3.45 and 3.46 can be written

$$P_T = K_1|E_f|\sin(\delta_0 + \Delta\delta) = P_{T0} + \Delta P_T \tag{3.47}$$
$$Q_T = K_1|E_f|\cos(\delta_0 + \Delta\delta) - K_2 = Q_{T0} + \Delta Q_T \tag{3.48}$$

where P_{T0} and Q_{T0} are values of P_T and Q_T at δ_0.

The normal operating value of δ is usually small. The variations in sine and cosine functions for small changes of δ are illustrated in Figure 3.30. The cosine function in the normal operating range of δ does not change very much. The value of Q_T can be accurately approximated as

$$\begin{aligned} Q_T &= K_1|E_f|\cos(\delta_0 + \Delta\delta) - K_2 \\ &= K_1|E_f|\cos\delta_0 - K_2 \\ &= Q_{T0} \end{aligned} \tag{3.49}$$

For changes in δ in the normal operating range, the value of generated reactive power is relatively unchanged. The sine function, however, does change significantly with δ. Therefore, P_T will change as δ changes.

If $|E_f|$ is changed an incremental amount, Equations 3.45 and 3.46 become

$$\begin{aligned} P_T &= K_1(E_{f0} + \Delta E_f)\sin\delta \\ &= P_{T0} + \Delta P_T \end{aligned} \tag{3.50}$$

$$\begin{aligned} Q_T &= K_1(E_{f0} + \Delta E_f)\cos\delta - K_2 \\ &= Q_{T0} + \Delta Q_T \end{aligned} \tag{3.51}$$

where E_{f0} and ΔE_f are magnitudes only.

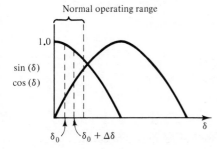

Figure 3.30 Variation in sine and cosine functions.

Expanding these equations, the values of ΔP_T and ΔQ_T are found:

$$\Delta P_T = K_1 \Delta E_f \sin \delta \qquad (3.52)$$

$$\Delta Q_T = K_1 \Delta E_f \cos \delta \qquad (3.53)$$

The value of ΔE_f is a small quantity, as is the value $\sin \delta$. Two small numbers multiplied together will result in ΔP_T being a relatively small value. However, the value of $\cos \delta$ is close to 1.0. Therefore, ΔQ_T will change significantly.

The amount of power being generated by a synchronous generator is a function of δ, and δ is controlled by the mechanical drive on the rotor. The amount of reactive power being generated by a synchronous generator is a function of $|E_f|$, and $|E_f|$ is controlled by the electrical excitation on the rotor. Both the rotor excitation and mechanical input are directly controllable by the generator operator. Since the amount of real and reactive power generated must equal the real and reactive power consumed, the flow of the real power and reactive power through the system can be controlled by varying the amount of each generated by the various generators in the system.

Another feature of synchronous generator operation concerns direction of reactive-power flow from the generator terminals. Example 3.12 shows that reactive-power flow out of the terminals of the generator of Example 3.11 is positive. Reactive-power flow is defined as positive when flowing from a generator into a load. A generator operating in this state is said to be overexcited. The opposite condition, called underexcited operation, occurs when reactive-power flow out of the terminals is negative. The following example problem illustrates these conditions.

EXAMPLE 3.14

The generator of Example 3.11 is connected to a load Z_L. Calculate the rotor current needed to hold the generator terminal voltage V_{ts} at 580 V line-to-neutral and also calculate S_T. Perform these calculations on the three loads given below where Z_L is the impedance in each of the three phases of the Y-connected load:

(a) $Z_L = 3 - j4 \, \Omega$
(b) $Z_L = 5 \, \Omega$
(c) $Z_L = 3 + j4 \, \Omega$

The equivalent circuit model needed to solve this problem is shown in Figure 3.31.

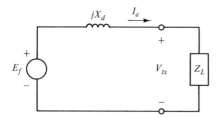

Figure 3.31 Equivalent circuit.

3.7 SYNCHRONOUS GENERATORS IN A POWER SYSTEM

Part A With V_{ts} choose as reference,

$$I_s = \frac{V_{ts}}{Z_L}$$

$$= \frac{580/0°}{5/-53.13°}$$

$$= 116/53.13° \text{ A}$$

Now solve for E_f:

$$E_f = V_{ts} + jX_d I_s$$
$$= 580/0° + (2.995/90°)(116/53.13°)$$
$$= 367/34.61° \text{ V}$$

The solution for I_r follows from Equation 3.40:

$$I_r = \frac{|E_f|}{\omega_s L_{sr}}$$

$$= \frac{367}{96.51}$$

$$= 3.803 \text{ A}$$

The three-phase power S_T is calculated as follows:

$$S_T = (3)(V_{ts})I_s^*$$
$$= (3)(580/0°)(116/-53.13°)$$
$$= 201,840/-53.13° \text{ VA}$$

or
$$S_T = 121,104 - j161,472 \text{ VA}$$

In this case the reactive power leaving the terminals of the generator is negative, and thus the generator is underexcited. This condition results from a capacitive load. Figure 3.32 shows a phasor diagram of E_f, V_{ts}, and I_s.

Part B Using the method of part A solve for I_r and S_T when $Z_L = 5 \, \Omega$.

$$I_s = \frac{580/0°}{5/0°}$$
$$= 116/0° \text{ A}$$

$$E_f = 580/0° + (2.995/90°)(116/0°)$$
$$= 676.1/30.92° \text{ V}$$

$$I_r = \frac{676.1}{96.51}$$

$$= 7 \text{ A}$$

$$S_T = (3)(580/0°)(116/0°)$$
$$= 201,840/0° \text{ VA}$$

or
$$S_T = 201,840 + j0 \text{ VA}$$

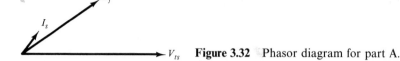

Figure 3.32 Phasor diagram for part A.

In this case the reactive power leaving the terminals of the generator is 0. This condition results from a resistive load. Figure 3.33 shows a phasor diagram for E_f, V_{ts}, and I_s.

Part C Using the method of part A, solve for I_R and S_T when $Z_L = 3 + j4 \ \Omega$.

$$I_s = \frac{580\underline{/0°}}{5\underline{/53.13°}}$$
$$= 116\underline{/-53.13°} \ \text{A}$$
$$E_f = 580\underline{/0°} + (2.995\underline{/90°})(116\underline{/-53.13°})$$
$$= 882.9\underline{/13.66°} \ \text{V}$$
$$I_R = \frac{882.9}{96.51}$$
$$= 9.148 \ \text{A}$$
$$S_T = (3)(580\underline{/0°})(116\underline{/53.13°})$$
$$= 201{,}840\underline{/53.13°} \ \text{VA}$$

or $\quad S_T = 121{,}104 + j161{,}472 \ \text{VA}$

In this case the reactive power leaving the terminals of the generator is positive, and thus the generator is overexcited. This condition results from an inductive load. Figure 3.34 shows a phasor diagram for part C.

■■

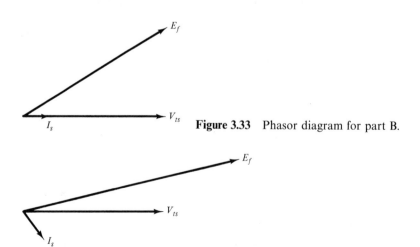

Figure 3.33 Phasor diagram for part B.

Figure 3.34 Phasor diagram for part C.

3.8 SUMMARY

The steady-state operation of a synchronous generator consists primarily of two phenomena. The interaction between the rotor and rotating stator MMFs result in a torque acting on the rotor. In steady-state operation the angles between the MMFs are constant, and therefore the torque is constant. The result of this interaction is the conversion of mechanical energy to electric energy.

The flux linking the stator windings induces a voltage on the stator coils. This voltage helps maintain the voltage level of the system. In addition, if the synchronous reactance of a machine is known, then the steady-state operating characteristics of a machine becomes a simple circuit problem of a reactance between two easily calculated voltages.

3.9 PROBLEMS

3.1. Consider again the rotating coil of Example 3.1. Calculate and plot the voltage induced on the coil for two revolutions for the following conditions:
   (a) Applied magnetic field B rotates in the same direction as the coil at a speed of 1.5 rad/s.
   (b) Applied magnetic field B rotates in the same direction as the coil at a speed of 2.0 rad/s.

3.2. Given the synchronous machine shown in Figure 3.35, plot the MMF of phases a, b, and c.

3.3. Given the synchronous machine shown in Figure 3.35, show that the MMF of phase a is a closer approximation to a cosine wave than the machine shown in Figure 3.10. (*Hint:* Use Fourier series analysis to analyze the harmonic content of each MMF.) Let N_s of Figure 3.10 equal $4N_s$ of Figure 3.35; that is, let both MMFs have the same maximum magnitude.

3.4. Calculate the expression for the stator MMF, F_s, of the synchronous machine shown in Figure 3.35.

3.5. Calculate the expression of the resultant MMF, R, of the synchronous machine shown in Figure 3.35. Assume that the rotor MMF F_R has the expression

$$F_R = \frac{4N_r i_r}{\pi} \cos(\psi - \theta)$$

where $\theta = 377t + 0.7\pi$ rad.

3.6. The coils of phase a of the machine shown in Figure 3.35 are shown in plan view in Figure 3.36. Note that phase a has four coils connected in series. Calculate the voltage induced on phase a at $t = 0.1$ s and with rotor excitation as given in Problem 3.5.

3.7. Calculate the values of L_s and L_{sr} for the machine of Figure 3.35. Use the rotor excitation conditions of Problem 3.5.

3.8. Calculate the mechanical torque applied to the machine of Figure 3.35. Use the rotor excitation conditions of Problem 3.5.

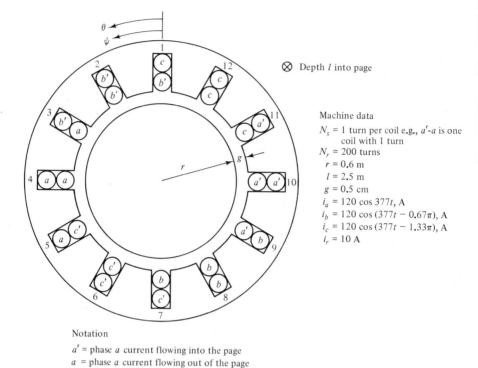

⊗ Depth *l* into page

Machine data
N_s = 1 turn per coil e.g., a'-a is one coil with 1 turn
N_r = 200 turns
r = 0.6 m
l = 2.5 m
g = 0.5 cm
i_a = 120 cos 377t, A
i_b = 120 cos (377t − 0.67π), A
i_c = 120 cos (377t − 1.33π), A
i_r = 10 A

Notation

a' = phase *a* current flowing into the page
a = phase *a* current flowing out of the page
other currents follow the same pattern

Figure 3.35 Synchronous machine.

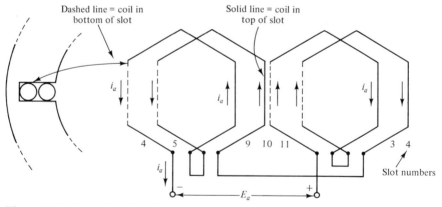

Figure 3.36 Plan view of coils in phase *a*.

3.9. Calculate the synchronous speed of the following generator types: two pole, four pole, six pole, eight pole, and ten pole.

3.10. Given the synchronous machine of Figure 3.37, plot the MMF of phase *a* with respect to ψ. How many poles does this machine have?

3.11. Calculate L_m, L_d, v_f, and v_s for the synchronous machine of Figure 3.35. Assume $L_l = 3 \times 10^{-4}$ H.

3.9 PROBLEMS

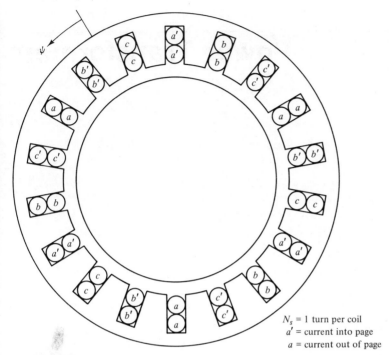

Figure 3.37 Synchronous machine.

3.12. A three-phase synchronous generator has a field excitation voltage of 1000 V and $X_d = 1.5\ \Omega$. A three-phase load of $10 + j15\ \Omega$ is connected to its terminals. Calculate the following quantities:
(a) V_{ts} (b) I_s
(c) S (d) power factor

3.13. Repeat Problem 3.12 for a load of $10 - j18\ \Omega$.

3.14. Repeat Problem 3.13 for a load of $20\ \Omega$.

3.15. For the conditions of Problem 3.12, calculate the value of $|E_f|$ needed to hold $|V_{ts}|$ at 1000 V.

3.16. Repeat Problem 3.15 for the conditions of Problem 3.13.

3.17. Repeat Problem 3.15 for the conditions of Problem 3.14.

3.18. A three-phase synchronous generator is operating with E_f at $1000\underline{/0°}$ and a load of $5 + j5\ \Omega$. Calculate S and the generator's pullout power. The value of X_d is $j1.0\ \Omega$.

3.19. A synchronous generator is operating with $E_f = 1500\underline{/0°}$ V, $V_{ts} = 1400\underline{/-80°}$ V, and $X_d = j2.0\ \Omega$. Calculate the three-phase power leaving its terminals and the value of the load connected to its terminals in ohms.

3.20. State the excitation conditions of the machine in Problem 3.12 (i.e., over-, normally, or underexcited).

3.21. State the excitation conditions of the machine in Problem 3.13.

3.22. State the excitation conditions of the machine in Problem 3.14.

Chapter 4

Power Transformers

4.1 INTRODUCTION

After electric power is generated by synchronous generators, it usually flows out into the system through power transformers. Figure 4.1 shows the relative position of these devices with respect to the generators and transmission system. The function of a transformer at this point in the power system is to step the terminal voltage of the generator up to the voltage of the transmission system. The reason for stepping up the voltage is to reduce I^2r losses within the transmission system. The three-phase power that is transmitted on a transmission line is

$$P = \sqrt{3} V_{ll} I_l \cos \theta \qquad (4.1)$$

In Equation 4.1, P is the power from the sending terminal, V_{ll} is the magnitude of the line-to-line voltage at that terminal, I_l is the magnitude of the line

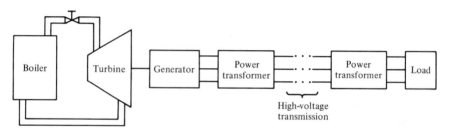

Figure 4.1 Placement of power transformer between the synchronous generator and high-voltage transmission system.

4.1 INTRODUCTION

Figure 4.2 Three-phase step-up transformer at a generating station.

current at the sending terminal, and θ is the angle between the line-to-neutral voltage and line current. If the power factor and power being transmitted remain constant, raising V_{ll} will reduce I_l. The line current will have an inverse relation with the line-to-line voltage. For example, doubling the voltage will decrease the current by half, with the same power being transmitted. The $I^2 r$ losses on the transmission line, however, will be decreased to one-fourth of their previous value. Therefore, for line loss considerations and reduction of voltage drop along the line, higher transmission voltages are desirable.

Figure 4.2 shows an example of a three-phase, step-up transformer at a generating station. The building behind the transformer houses the synchronous generator which has a terminal voltage of approximately 12 kV at 330 MW. The transformer increases this voltage to 345 kV, which appears at the terminals of the large bushings on top of the transformer.

Figures 4.3 and 4.4 show other kinds of power transformers located in a high-voltage substation. Figure 4.3 show a three-phase, 345/765-kV, 1000-MVA transformer bank. Figure 4.4 shows a spare transformer (on the left) in the same station. The device to the right of the transformer is a spare 765-kV shunt reactor. A shunt reactor serves the purpose of current limiting during line-to-ground short circuits on a transmission line and may also be used for reactive-power compensation.

Figure 4.1 also shows power transformers between the transmission system and the customer electric load. These transformers reduce the voltages of the transmission system down to levels usable by customers. Usually, the lowest

Figure 4.3 345/765-kV, 1000-MVA three-phase transformer bank.

Figure 4.4 Spare 345/765-kV, 333-MVA transformer (on the left) and spare 765-kV shunt reactor (on the right).

4.1 INTRODUCTION

Figure 4.5 Single-phase pole-mounted transformer, high-voltage bushings.

customer voltage is in residential areas where single-phase, center-tapped transformers supply 240 and 120 V to houses and small businesses.

Figures 4.5 and 4.6 show a single-phase pole-mounted transformer. Figure 4.5 shows the higher-voltage (usually several thousand volts) bushings on top of the transformer connected to a distribution transmission line through a fuse and voltage surge arrestor. The fuse protects the transformer from short circuits. The surge arrestor protects the transformer from high-voltage surges caused by nearby lightning strokes.

Figure 4.6 shows a different view of the same transformer. The low-voltage single-phase power appears at the three lugs on the side of the transformer. The two outer lugs carry 240 V between them, and the center lug carries 120 V between itself and the outer lugs. Four separate houses are served by this transformer, as can be seen by the four sets of wires leading away from the pole.

Figure 4.7 shows a residential pad-mounted transformer in the foreground. The metal case encloses the transformer, and the lines connected to it are buried in the ground. Underground distribution such as this is more common among newer housing subdivisions.

Figure 4.6 Single-phase pole-mounted transformer, low-voltage service.

Figure 4.7 Single-phase pad-mounted transformer.

4.2 SINGLE-PHASE TRANSFORMERS

Transformers are static devices that step voltages up or down. For a single-phase application, this voltage transformation is achieved by two coils wrapped around a common iron core. These two coils are called the primary and the secondary. The primary coil is connected to the electrical source, while the secondary coil is connected to the load. A simplified illustration of a power transformer with a voltage source on the primary and a load on the secondary is shown in Figure 4.8. N_1 and N_2 are the number of turns in each coil. The flux, ϕ_{11}, is the flux linking coil 1 only, and ϕ_{22} is the flux linking coil 2 only. The flux ϕ_{11} is called the effective leakage flux of coil 1 and is a function of i_1 but not i_2. The flux ϕ_{12} links both coils 1 and 2 and is called the mutual flux. The mutual flux is a function of both i_1 and i_2.

The voltage across any ideal coil is equal to the time rate of change of the flux linking the coil. Using this definition, the voltage across the primary of the transformer in Figure 4.8 is

$$v_1 = i_1 r_1 + \frac{d\lambda_1}{dt} \tag{4.2}$$

The quantity λ_1 is the total flux linkage for coil 1. The resistance, r_1, is present because the coils on a transformer are not ideal and a small resistance is associated with each of them. Recall from Chapter 2 that $\lambda = N\phi$; thus, Equation 4.2 can be written as

$$v_1 = i_1 r_1 + \frac{dN_1(\phi_{11} + \phi_{12})}{dt} \tag{4.3}$$

Since ϕ_{11} is a function of i_1 only, an effective leakage inductance may be substituted for $dN_1 \phi_{11}/dt$.

$$v_1 = i_1 r_1 + L_{11}\frac{di_1}{dt} + \frac{dN_1 \phi_{12}}{dt} \tag{4.4}$$

A similar analysis can be done for coil 2:

$$v_2 = -i_2 r_2 + \frac{d\lambda_2}{dt} \tag{4.5}$$

$$v_2 = -i_2 r_2 + \frac{dN_2(-\phi_{22} + \phi_{12})}{dt} \tag{4.6}$$

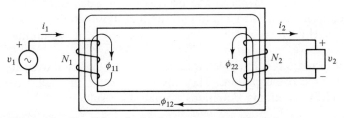

Figure 4.8 Simplified single-phase power transformer.

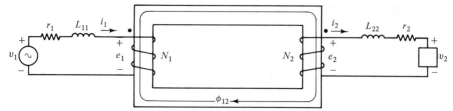

Figure 4.9 Single-phase transformer with leakage inductances.

$$v_2 = -i_2 r_2 - L_{22}\frac{di_2}{dt} + \frac{dN_2 \phi_{12}}{dt} \quad (4.7)$$

The negative sign on the leakage flux appears because a positive current i_2 creates a leakage flux with a positive direction that, according to the right-hand rule, is opposite to the direction of the mutual flux. Figure 4.9 illustrates Equations 4.4 and 4.7.

4.2.1 Voltage Transformation

In Figure 4.9 e_1 and e_2 are voltages induced on the coils due to changes in the mutual flux only. The sign references for these voltages are determined by Lenz's law. Lenz's law states that a voltage induced on a coil will tend to produce a current in a direction so as to oppose the change in flux. As an example, if ϕ_{12} in Figure 4.9 were increasing, a voltage would be induced on both coils to oppose that increase. Since a positive i_2 would produce a flux in the opposite direction of ϕ_{12}, e_2 would increase in value in order to increase i_2. Therefore,

$$e_2 = \frac{dN_2 \phi_{12}}{dt} \quad (4.8)$$

Since a positive i_1 would produce a flux in the same direction as ϕ_{12}, e_1 must change so as to decrease i_1. Assuming the source voltage v_1 to be constant, increasing e_1 would decrease the voltage drop across the primary coil resistance and leaking inductance and would decrease i_1.

$$e_1 = \frac{dN_1 \phi_{12}}{dt} \quad (4.9)$$

Both e_1 and e_2 increase as ϕ_{12} increases and decrease as ϕ_{12} decreases. Dots are placed at the top terminal of each coil to indicate that those terminals have the same relative polarities for voltages induced from changes in the mutual flux.

N_1 and N_2 of Equations 4.8 and 4.9 are constants and can be moved to the left side of their respective equations. This leaves $d\phi_{12}/dt$ on the right side of both equations. Then combining Equations 4.8 and 4.9 results in

$$\frac{e_1}{N_1} = \frac{e_2}{N_2} \quad (4.10)$$

4.2 SINGLE-PHASE TRANSFORMERS

Equation 4.10 implies that the voltage per turn induced by the mutual flux is the same for both coils. If the transformer is excited by a sinusoidal source, Equation 4.10 can be rearranged and written in phasor form:

$$\frac{E_1}{E_2} = \frac{N_1}{N_2} = a \qquad (4.11)$$

The constant a is the ratio of transformation. In most cases the voltage drops across the coils' resistances and leakage inductances are very small compared to the phasor voltages V_1, V_2, E_1, and E_2. Therefore, the constant a is often approximated as the ratio of primary terminal voltage to secondary terminal voltage:

$$a \approx \frac{V_1}{V_2} \qquad (4.12)$$

4.2.2 Current Transformation

The path for the mutual flux in Figure 4.9 is the iron core. The reluctance of the core will be included as $\mathcal{R}$. The MMF which produces the mutual flux is the summation of the primary and secondary MMFs, F_1 and F_2. If once again sinusoidal quantities are assumed, then

$$\begin{aligned} F_{12} &= F_1 + F_2 \\ &= \phi_{12}\mathcal{R} \end{aligned} \qquad (4.13)$$

The current entering the dotted terminal of a coil is easily shown to produce a positive MMF. Therefore, for the references shown in Figure 4.9

$$F_1 = N_1 I_1 \qquad (4.14)$$

$$F_2 = N_2(-I_2) \qquad (4.15)$$

$$\begin{aligned} F_{12} &= N_1 I_1 - N_2 I_2 \\ &= \phi_{12}\mathcal{R} \end{aligned} \qquad (4.16)$$

Solving for I_1 yields

$$\frac{\phi_{12}\mathcal{R}}{N_1} + \frac{N_2}{N_1} I_2 = I_1 \qquad (4.17)$$

The first term of Equation 4.17, $\phi_{12}\mathcal{R}/N_1$, is called the exciting current. It is the current necessary to produce the mutual flux ϕ_{12}. Even with the secondary open-circuited, a changing mutual flux must exist in order to have a secondary voltage. Therefore, an exciting current must always be present. The second term of Equation 4.17 is the load component of the primary current. As the transformer is loaded, the load component becomes much larger than the exciting current. Therefore, under loaded conditions Equation 4.17 reduces to

$$I_1 \approx \frac{N_2}{N_1} I_2 = \frac{I_2}{a} \qquad (4.18)$$

Figure 4.10 illustrates the transformer with phasor quantities.

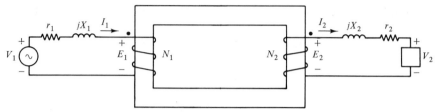

Figure 4.10 Single-phase transformer with phasor voltages and currents.

4.2.3 Exciting Current

The exciting current I_ϕ, as described in the previous section, is composed of two components, magnetizing current I_m and current-supplying core losses I_c:

$$I_\phi = I_m + I_c \tag{4.19}$$

The current-supplying core losses is also composed of two components. One part supplies the eddy current losses, and the other part supplies hysteresis losses.

Magnetizing Current From Equation 4.9 we recall that the voltage e_1 is a function of the time rate of change of the flux ϕ_{12}:

$$e_1 = N_1 \frac{d\phi_{12}}{dt} \tag{4.20}$$

In phasor form this equation can be written

$$E_1 = j\omega N_1 \phi_{12} \tag{4.21}$$

The voltage E_1 leads the flux ϕ_{12} by 90°. Since $N\phi = \lambda = Li$, Equation 4.21 can be written

$$\begin{aligned} E_1 &= j\omega L_m I_m \\ &= jX_m I_m \end{aligned} \tag{4.22}$$

I_m is the magnetizing current that produces the mutual flux, and X_m is the magnetizing reactance of the transformer.

Eddy Current Losses One of the basic principles behind the operation of transformers is that a voltage is induced around a contour enclosing a changing magnetic field. As the flux varies through the turns of the secondary coil, a voltage is produced on the turns. However, the flux is also varying everywhere in the iron core, and voltages are created around all the flux lines everywhere in the core. These voltages result in currents flowing in the core. These currents flow in closed paths perpendicular to the flux path. Since the iron core has a resistance, $I^2 r$ losses due to these eddy currents will occur. These losses are called eddy current losses and are a source of real-power loss in transformers. Thus, the current associated with them is in phase with the voltage producing the mutual flux.

4.2 SINGLE-PHASE TRANSFORMERS

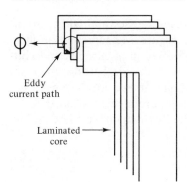

Figure 4.11 Eddy currents in a laminated core.

The magnitude of eddy currents can be diminished by laminating the iron core. Laminating the core means several thin layers of iron are used to form the core. The layers are assembled so that the eddy currents will have to flow across the junction of different layers, as shown in Figure 4.11. The junctions increase the resistance of the eddy current path and, therefore, decrease the magnitude of the eddy currents. Ferromagnetic alloys with higher resistivities than pure iron can also be used to limit eddy currents.

Hysteresis Losses From the classical theory of atomic structure, electrons tend to travel around the nucleus. This flow of negative charge results by definition in a current flow. Since these electrons travel in closed paths, a magnetic orientation is set up on each atom. If an external magnetic field is applied to these atoms, they will tend to align their magnetic orientation with that of the external field.

When a 60-Hz excitation is applied to an iron core transformer, the magnetic field reverses direction 120 times per second. The atoms in the core of the transformer must realign themselves with the field every time it changes direction. The energy used in this realignment is called hysteresis loss and is a source of real-power loss. Thus, the current associated with hysteresis loss is in phase with the voltage producing the mutual flux.

Figure 4.12 shows the transformer with magnetizing current and core losses represented. I'_1 is the load component of the primary current and r_c is

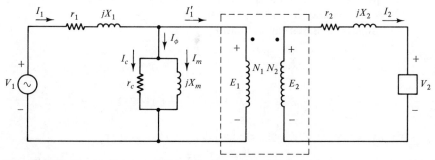

Figure 4.12 Single-phase transformer with magnetizing current and core loss representation.

the shunt resistance, which represents the hysteresis losses and eddy current losses. The coils enclosed in the dotted lines in Figure 4.12 represent an ideal transformer. No losses occur in the ideal transformation of currents and voltages inside the box. All nonideal characteristics of the ideal transformer are accounted for by elements external to these coils. In the ideal transformer winding resistance, core losses, and magnetizing current are assumed to be negligible. In addition, it is assumed that all the flux remains in the core and links both windings. That is,

$$I'_1 = \frac{I_2}{a} \tag{4.23}$$

$$E_2 = \frac{E_1}{a} \tag{4.24}$$

and
$$E_1(I'_1)^* = E_2 I_2^* \tag{4.25}$$

4.2.4 Measuring r_c and X_m

With reference to Figure 4.12, if the secondary of the transformer is open-circuited, then $I_2 = I'_1 = 0$. Therefore, the primary current is equal to the exciting current. Under these conditions the primary current is usually in the range of 2 to 5 percent of full-load current. This lower value of current indicates that the shunt impedances, r_c and X_m, are very large compared to the primary coil's resistance and leakage reactance.

With the secondary opened, the primary voltage is across a series combination of the primary coil's resistance and leakage reactance and the shunt impedances. Because of the relative magnitude of these two sets of impedances, that series combination can be approximated as the value of the shunt impedances. Therefore, in an open-circuit test, the real power being consumed is assumed equal to the core losses, and the reactive power being consumed is assumed to be used entirely to magnetize the core.

EXAMPLE 4.1

A transformer, which is rated for 50 kVA and 4000 V on the primary, has open-circuit test readings from the primary side of 4 kV, 200 W, and 0.35 A. Determine the values of r_c and X_m.

Since r_c is much larger than r_1,

$$P = \frac{V^2}{r_c}$$

$$r_c = \frac{V^2}{P} = \frac{(4 \times 10^3)^2}{200} = 80{,}000 \ \Omega$$

Also, since X_m is much larger than X_1,

$$Q = \frac{V^2}{X_m}$$

4.2 SINGLE-PHASE TRANSFORMERS

The value of Q must be calculated from the given data:

$$|S| = VI = (4000)(0.35) = 1400 \text{ VA}$$

$$\text{pf} = \frac{P}{|S|} = \frac{200}{1400} = 0.143 \text{ lagging}$$

$$\text{pf angle } \theta = \cos^{-1} 0.143 = 81.8°$$

$$Q = VI \sin \theta = (4000)(0.35) \sin 81.8° = 1386 \text{ VAR}$$

$$X_m = \frac{V^2}{Q} = \frac{(4 \times 10^3)^2}{1386} = 11{,}544 \text{ }\Omega \quad \blacksquare\blacksquare$$

4.2.5 Referring Impedances

Consider an ideal transformer with a load connected to it as illustrated in Figure 4.13. Kirchhoff's voltage law applied around the secondary circuit gives

$$E_2 = I_2 Z_2 \tag{4.26}$$

Equation 4.26 can be rearranged to give

$$Z_2 = \frac{E_2}{I_2} \tag{4.27}$$

Substituting for E_2 and I_2 yields

$$Z_2 = \frac{E_1/a}{aI_1} \tag{4.28}$$

or

$$a^2 Z_2 = \frac{E_1}{I_1} = Z_1 \tag{4.29}$$

Equation 4.29 indicates that, as far as the primary is concerned, a load on the secondary of a transformer can be represented as an equivalent impedance on the primary. The equivalent impedance Z_1 would have the value of $a^2 Z_2$. Therefore, the circuit of Figure 4.13 is equivalent to the circuit of Figure 4.14.

Without the ability to refer impedances across an ideal transformer, analysis of most circuits containing transformers would be difficult. For instance, in Figure 4.12 if the load on the secondary is $r_L + jX_L$ and the only voltage that is known is V_1, the solution of the other system voltages and currents becomes a complex problem. However, if the impedance $r_2 + r_L + j(X_2 + X_L)$ is

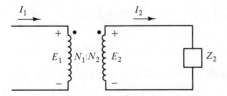

Figure 4.13 Single-phase ideal transformer.

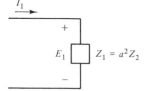

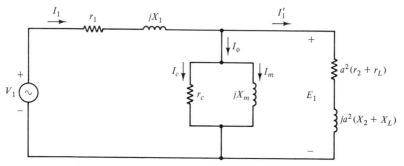

Figure 4.14 Secondary impedance referred to the primary of ideal transformer.

Figure 4.15 Secondary impedance referred to the primary of transformer equivalent circuit.

referred to the primary, the equivalent circuit appears as shown in Figure 4.15. All primary quantities can be easily calculated using this circuit. Secondary voltage and currents can then be determined using the ratio of transformation.

Impedances can also be referred from the primary to the secondary by dividing the primary impedance by a^2.

EXAMPLE 4.2

For the transformer in Example 4.1, the coil resistances and linkage reactances are $r_1 = 1.2 \, \Omega$, $X_1 = 4.0 \, \Omega$, $r_2 = 0.012 \, \Omega$, and $X_2 = 0.04 \, \Omega$. The ratio of transformation for the transformer is 10.0. If a load of $Z = 4.0 + j2.0 \, \Omega$ is connected to the secondary and the primary is supplied by 4000 V, find the primary and secondary currents and the secondary voltage.

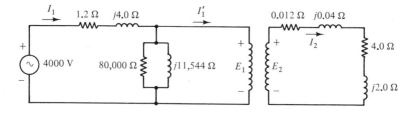

The secondary impedances must be referred to the primary side.

$$a^2(r_2 + jX_2 + 4.0 + j2.0) = (10)^2(4.012 + j2.04)$$
$$= 401.2 + j204 \, \Omega$$

4.2 SINGLE-PHASE TRANSFORMERS

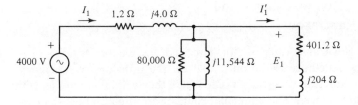

This circuit can be easily reduced:

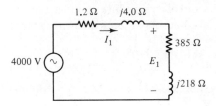

Choose V_1 as reference and solve for I_1.

$$I_1 = \frac{4000 + j0}{386.2 + j222}$$
$$= 8.98\underline{/-29.9°} \text{ A}$$

Using voltage division,

$$E_1 = \frac{385 + j218}{386.2 + j222}(4000 + j0)$$
$$= 3973\underline{/-0.4°} \text{ V}$$

E_2 can now be found:

$$E_2 = \frac{E_1}{a}$$
$$= \frac{3973\underline{/-0.4°}}{10}$$
$$= 397.3\underline{/-0.4°} \text{ V}$$

Now I_2 can be found from the original circuit:

$$I_2 = \frac{E_2}{Z_2 + Z}$$
$$= \frac{397.3 + j2.774}{4.012 + j2.04}$$
$$= 88.3\underline{/-27.4°} \text{ A}$$

Finally, the secondary voltage can be calculated:

$$V_2 = I_2 Z$$
$$= (78.39 - j40.64)(4 + j2)$$
$$= 395\underline{/-0.8°} \text{ V}$$

■■

4.2.6 Measuring Coil Resistance and Leakage Reactance

If the secondary of a transformer is short-circuited, the secondary coil's resistance and leakage reactance is the only load on the transformer. These impedances can be referred to the primary side where they will be in parallel with the shunt core impedances. The shunt impedances are much larger than the secondary coil's resistance and leakage reactance referred to the primary. Therefore, the parallel combination of these two sets of impedances can be approximated as the referred secondary coil values. This procedure is illustrated in Figure 4.16. The primary side voltage is placed across an equivalent resistance and reactance with the values

$$r_{eq} = r_1 + a^2 r_2 \tag{4.30}$$

$$X_{eq} = X_1 + a^2 X_2 \tag{4.31}$$

By measuring the voltage, current, and power at the primary terminals, r_{eq} and X_{eq} can be determined. Since the values of r_{eq} and X_{eq} are relatively low, a voltage significantly less than normal operating or rated voltage is used for the short-circuit test so that normal operating current flows in the primary circuit. This test condition results in a value of X_{eq} at a flux associated with normal operating current.

The value of r_{eq} may be divided into r_1 and r_2 in the following manner. Measurements of the dc resistance of the primary and secondary coils, r_{1dc} and

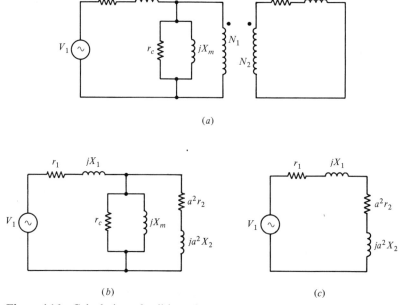

Figure 4.16 Calculation of coil impedances.

4.2 SINGLE-PHASE TRANSFORMERS

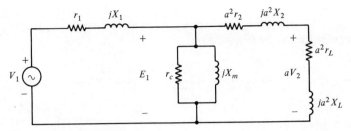

Figure 4.17 Equivalent circuit.

r_{2dc}, are made. These resistances will be less than their corresponding 60-Hz values because of additional losses and increased resistances associated with ac operation, such as eddy currents, skin effect, and induced current flow in nearby conducting material. However, the ratio of r_1 to r_2 can be assumed to be equal to the ratio of r_{1dc} to r_{2dc}. Therefore,

$$\frac{r_1}{a^2 r_2} = \frac{r_{1dc}}{a^2 r_{2dc}} \tag{4.32}$$

and

$$r_1 = \left(\frac{r_{1dc}}{r_{1dc} + a^2 r_{2dc}}\right) r_{eq} \tag{4.33}$$

The same ratio may be assumed for X_1 and X_2. In most cases r_1 is very close in value to $a^2 r_2$, and X_1 is almost equal to $a^2 X_2$.

4.2.7 Approximate Equivalent Circuits

The equivalent circuit used so far in this chapter for a single-phase transformer is shown in Figure 4.17 with all impedances referred to the primary. The impedances r_L and X_L represent the load on the transformer, and V_1 is the electric source. Recall that the impedances r_1 and X_1 usually have relatively low values. Therefore, the voltage drop across them is small, and E_1 is approximately equal to V_1. With this approximation the equivalent circuit can be drawn as shown in Figure 4.18. This approximate equivalent circuit indicates that the exciting current is not a function of the load on the transformer and makes calculations much easier.

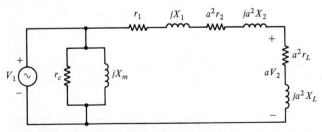

Figure 4.18 Approximate equivalent circuit.

EXAMPLE 4.3

Repeat Example 4.2 using the approximate equivalent circuit. The impedance diagram would be

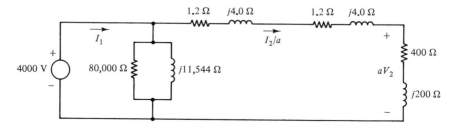

With V_1 as reference, I_2 can be calculated:

$$\frac{I_2}{a} = \frac{V_1}{(1.2 + 1.2 + 400) + j(4 + 4 + 200)}$$

$$I_2 = 88.3 \underline{/-27.3°} \text{ A}$$

Next V_2 is calculated:

$$aV_2 = \frac{I_2}{a}(400 + j200)$$

$$V_2 = 395 \underline{/-0.7°} \text{ V}$$

Finally I_1 can be determined:

$$I_1 = \frac{V_1}{80,000} + \frac{V_1}{j11,544} + \frac{I_2}{a}$$

$$I_1 = 9.04 \underline{/-29.1°} \text{ A}$$

These answers correspond closely with those found in Example 4.2. Less than 1 percent error was introduced in the magnitude of I_1, I_2, or V_2 by using the approximate equivalent circuit. ∎

In a previous section exciting current was shown to be a very small percentage of the primary current, under normal operating conditions. The equivalent circuit could be greatly simplified by neglecting the exciting current entirely. The circuit that results is called the series impedance approximate equivalent circuit and is shown in Figure 4.19. $I_1 = I_2/a$ in this model. The

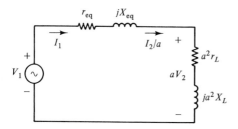

Figure 4.19 Series impedance approximate equivalent circuit.

4.2 SINGLE-PHASE TRANSFORMERS

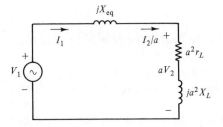

Figure 4.20 Series reactance approximate equivalent circuit.

values of r_{eq} and X_{eq} are the values measured directly from the short-circuit test. They do not have to be separated into r_1, r_2, X_1, and X_2 as shown in Equations 4.30 and 4.31. The calculated values of V_2 and I_2 will be very close to the values calculated in Example 4.2, and the error in the magnitude of I_1 will be only a few percent. The ease of calculation with this equivalent circuit more than compensates for the small error that is introduced.

Since the leakage reactance of a transformer is usually significantly larger than the coil resistances, the resistance in the equivalent circuit of Figure 4.19 is often neglected. This gives the series reactance approximate equivalent circuit, which is illustrated in Figure 4.20.

EXAMPLE 4.4

Repeat Example 4.2 using the series reactance approximate equivalent circuit.

The impedance diagram is

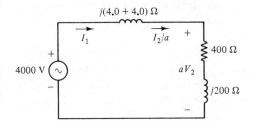

With V_1 as reference, I_1 and I_2 can be calculated directly:

$$I_1 = \frac{V_1}{400 + j(4 + 4 + 200)} = 8.87\underline{/-27.5°} \text{ A}$$

$$I_2 = aI_1 = 88.7\underline{/-27.5°} \text{ A}$$

Now V_2 can be found:

$$V_2 = \frac{I_1}{a}(400 + j200) = 397\underline{/-0.9°} \text{ V}$$

Again the error in the magnitudes and angles of I_1, I_2, and V_2 is only a few percent. The calculations for this model become almost trivial. For these reasons the series reactance approximate equivalent circuit is used in the analysis of most power system problems. ∎

4.2.8 Physical Considerations

Unlike in the representation in Figure 4.8, the windings of a transformer are not separated on opposite sides of the core. Usually one-half of the primary windings and one-half of the secondary windings are layered on top of each other on each side of the core. This arrangement minimizes the leakage flux in both coils.

The losses of a transformer are dissipated as heat. Excessive heat can damage insulation and cause the transformer to fail. Therefore, the heat must be removed from the vicinity of the coils and core. Heat removal is done by placing the coils in an oil-filled tank. The oil acts as a medium for conducting the heat away from the core and coils to the outside air. For larger transformers the oil is forced from the core into radiators before circulating back over the core. The radiators help dissipate more heat from the oil to the air. Larger transformers are often fitted with fans to force more cooling air over the radiators (see Figures 4.3 and 4.4).

Since losses are dissipated as heat, and heat is a limiting factor in the life of a transformer, losses must be held to some value which will not result in damage to the transformer. As was discussed in the previous section, core losses are almost constant for a given terminal voltage. Copper losses are dependent on the square of the magnitude of the current flowing in each coil. Therefore, at rated voltage, the losses would have some minimum core loss value and would increase as current increases. Since current magnitude determines the upper limit of transformer losses, transformers are rated by voltamperes and not by watts. If the load on a transformer were purely reactive, it could be required to withstand maximum losses while delivering no real power to its secondary.

Two quantities that help describe the operation of a transformer are its efficiency and voltage regulation. Efficiency is defined as power output divided by power input and is usually given as a percent quantity. Voltage regulation is defined as the change in the secondary voltage from no load to full load as a percentage of the full-load voltage. In equation form it is

$$\text{Regulation} = \frac{\text{secondary no-load voltage} - \text{secondary full-load voltage}}{\text{secondary full-load voltage}} \times 100$$

4.2.9 Multiwinding Transformers

Power transformation between more than two voltage levels in one transformer is not uncommon. For instance, at a substation where a 345/138-kV transformer is used, low-voltage power for security lighting may be needed. For these situations a third winding, called a tertiary winding, is often wound on the transformer as shown in Figure 4.21. With the correct number of turns, a desired voltage can be obtained. Since the same mutual flux flows through each coil, the voltage on the primary, secondary, and tertiary are all in phase. Their magnitudes are determined by the number of turns of each coil.

4.2 SINGLE-PHASE TRANSFORMERS

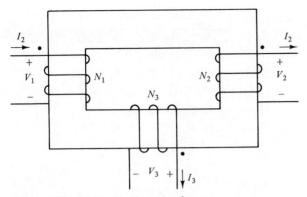

Figure 4.21 Three-winding transformer.

If Equations 4.16 and 4.17 are expanded to include a third coil and magnetizing current is neglected, we see that the MMF balance between coils must be maintained:

$$N_1 I_1 = N_2 I_2 + N_3 I_3 \tag{4.34}$$

EXAMPLE 4.5

Consider the transformer in Figure 4.21 to be ideal. The load on the tertiary is $5.0\ \Omega$. The load on the secondary is $15 + j5.0\ \Omega$. If the primary voltage is 1.0 kV and N_1, N_2, and N_3 are 1000, 400, and 100, respectively, find I_1.

With V_1 chosen as reference, the secondary and tertiary voltages are

$$V_2 = V_1 \frac{N_2}{N_1} = 400\underline{/0°}\ \text{V}$$

$$V_3 = V_1 \frac{N_3}{N_1} = 100\underline{/0°}\ \text{V}$$

I_2 and I_3 can now be calculated:

$$I_2 = \frac{V_2}{15 + j5} = 25.3\underline{/-18.4°}\ \text{A}$$

$$I_3 = \frac{V_3}{5} = 20.0\underline{/0°}\ \text{A}$$

Balancing MMFs gives

$$N_1 I_1 = N_2 I_2 + N_3 I_3$$

$$I_1 = \frac{(400)(25.3\underline{/-18.4°}) + (100)(20.0\underline{/0°})}{1000}$$

$$= 11.6 - j3.2 = 12.0\underline{/-15.4°}\ \text{A} \qquad \blacksquare\blacksquare$$

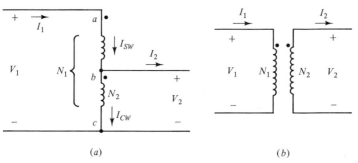

Figure 4.22 Autotransformer.

4.2.10 Autotransformers

So far only transformers whose coils are connected solely by magnetic coupling have been examined. Transformers can be constructed so that the primary and secondary coils are also electrically connected. This type of transformer is called an autotransformer and is compared to a two-winding transformer in Figure 4.22. The coil from points a to c of the autotransformer is the primary. The winding from a to b is called the series winding. The winding from b to c is called the common winding. The common winding makes up the secondary coil. Compare the autotransformer and the two-winding transformer. If the primary voltages and the number of turns on the primary coils are the same for both the autotransformer and the two-winding transformer, then the volts per turn will be the same. Since the same mutual flux passes through the secondary as the primary in both cases, the volts per turn will be the same for both secondaries. Therefore, if the number of turns on the secondary coil is the same for the autotransformer as the two-winding transformer, the secondary voltages will also be equal.

EXAMPLE 4.6

An ideal autotransformer with a turns ratio of 2:1 has a high side voltage of 12 kV. If the low-voltage side serves a single-phase load of 1200 kW at unity power factor, determine the current flow in the series and common windings.

The voltage on the secondary with the primary voltage chosen as reference is

$$V_2 = \frac{V_1}{2} = 6.0 \underline{/0°} \text{ kV}$$

The secondary current is

$$I_2^* = \frac{S_2}{V_2} = \frac{1200 \times 10^3 \underline{/0°}}{6 \times 10^3 \underline{/0°}} = 200 \underline{/0°} \text{ A}$$

4.2 SINGLE-PHASE TRANSFORMERS

Since the transformer is ideal, the primary current is

$$I_1 = \frac{I_2}{2} = 100\underline{/0°} \text{ A}$$

The series-winding current equals

$$I_{SW} = I_1 = 100\underline{/0°} \text{ A}$$

The common winding current is

$$I_{CW} = I_1 - I_2 = -100\underline{/0°} \text{ A}$$

The transformer currents can be represented on the autotransformer diagram as shown:

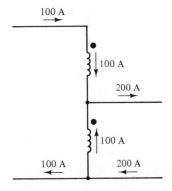

■■

Since a separate secondary coil is not used, and no coil carries the full secondary current, significant savings can be made in the copper used for the coils of an autotransformer. On the whole, autotransformers cost less, have less leakage flux, have fewer losses, and require less exciting current than two-winding transformers. As the turns ratio approaches 1, the advantage of autotransformers is the greatest. However, they have one major disadvantage. If the common winding were to open circuit, then full high-side voltage would appear on the low-side terminals, resulting in a safety hazard. This situation could not happen with a two-winding transformer.

4.2.11 Core Nonlinearities

In the development of the equivalent circuits for a transformer, a sinusoidal current was assumed to create a sinusoidal flux. A sinusoidal flux is necessary so that sinusoidal voltages will be induced on the coils. Recall that $B = \mu H$, that B is directly related to flux, and that H is directly related to current. Therefore, if a sinusoidal current creates a sinusoidal flux of the same frequency, then μ of the core must be a constant. Under normal operating conditions, however, the μ of an iron core is not constant. As the amount of flux in the core increases,

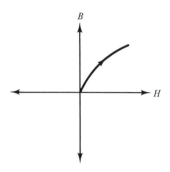

Figure 4.23 Increasing B and H from a deenergized core.

more current is required to cause the next increment of flux to flow than the previous increment of flux. Figure 4.23 illustrates this relationship starting from a deenergized core and increasing the flux to its maximum value. If H is decreased, B decreases, but not along the same path as it increased. As the magnetic field was building in the core, the atoms in the core tended to orient themselves magnetically with the field. As the current decreases and reverses in direction, some of these atoms tend to stay in their present orientation due to a magnetic friction. Therefore, some of the flux remains in the same orientation. The relationship between current and flux or H and B is, therefore, dependent upon the immediate past history of the core. This concept is illustrated in Figure 4.24, which differs from Figure 4.23 in that the core is not starting from a deenergized state. For each sinusoidal cycle of flux the closed path of Figure 4.24 will be traveled once. The area inside the loop is proportional to the hysteresis losses of the core per cycle.

The above discussion clearly shows that μ is not constant. Therefore, in order to maintain a sinusoidal flux, the current creating the flux must be nonsinusoidal. A harmonic analysis of transformer current reveals that it has the fundamental frequency plus all odd harmonics of the fundamental. The third harmonic is the largest and most troublesome. Most power systems are constructed using methods that control harmonics. Therefore, for most steady-state system studies the harmonics are neglected and the 60-cycle components are used with the linear equivalent circuits.

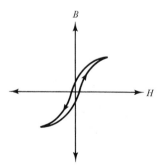

Figure 4.24 Relationship of B and H.

4.3 THREE-PHASE TRANSFORMER BANKS

Three-phase power can be transformed by connecting three single-phase transformers in one of four ways. These connections are called Y-Y, Δ-Δ, Δ-Y, and Y-Δ. These same connections can also be used in one three-phase transformer.

4.3.1 Y-Y Connection

If the transformers are connected as shown in Figure 4.25, the connection is called a Y-Y. Each transformer carries one-third of the power being transformed under balanced conditions and transforms line-to-neutral voltages and line currents. If the three transformers are considered ideal, the primary line-to-neutral voltages and line currents are transformed to the secondary such that the secondary voltages and currents are in phase with their corresponding primary quantities.

Some of the advantages of the Y-Y connection are that a neutral point is available for grounding purposes to both the high- and low-voltage sides, and under normal operation, each transformer needs to be insulated for only line-to-neutral voltage levels. However, as discussed in the previous section, in order for the secondary voltage to be sinusoidal, the exciting current must contain third harmonics. The presence of third-harmonic components can create some problems.

Since the fundamental of each line current is 120° out of phase with the other two fundamentals, the third harmonics of each line will be in phase:

$$I_{a1} = \sqrt{2}I_a \cos \omega t$$
$$I_{b1} = \sqrt{2}I_b \cos(\omega t - 120°)$$
$$I_{c1} = \sqrt{2}I_c \cos(\omega t + 120°) \quad (4.35)$$

$$I_{a3} = \sqrt{2}I_3 \cos 3\omega t$$
$$I_{b3} = \sqrt{2}I_3 \cos 3(\omega t - 120°)$$
$$= \sqrt{2}I_3 \cos(3\omega t - 360°)$$
$$= \sqrt{2}I_3 \cos 3\omega t$$

$$I_{c3} = \sqrt{2}I_3 \cos 3(\omega t + 120°)$$
$$= \sqrt{2}I_3 \cos(3\omega t + 360°)$$
$$= \sqrt{2}I_3 \cos 3\omega t \quad (4.36)$$

Therefore, the third-harmonic currents will not sum to zero at the neutral point under balanced conditions as the fundamental currents do. The third harmonics will sum algebraically, and that summation of current will flow in the neutral. Each line will carry its component of the third-harmonic into the power system thus making a Y-Y connection a source of harmonics in the power system.

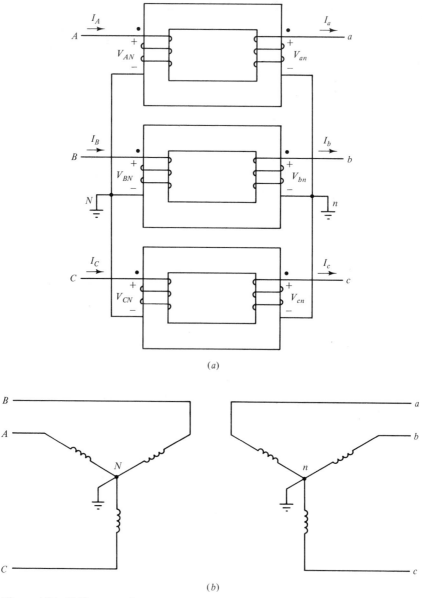

Figure 4.25 Y-Y connection.

4.3.2 Δ-Δ Connection

The second type of connection is the Δ-Δ connection. This connection is illustrated in Figure 4.26. Each transformer carries one-third of the power being transformed under balanced conditions. Line-to-line voltages and Δ phase currents are transformed by each transformer. The transformed secondary line-to-

4.3 THREE-PHASE TRANSFORMER BANKS

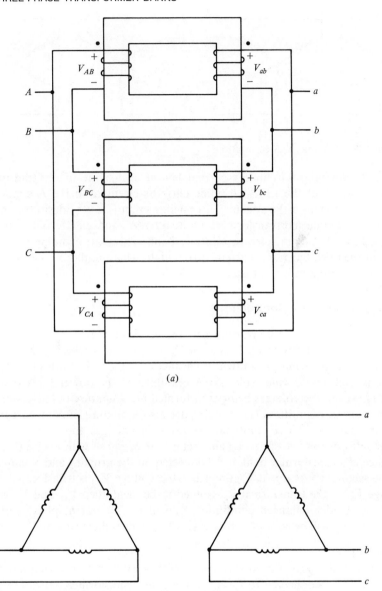

Figure 4.26 Δ-Δ connection.

line voltages and phase currents are in phase with their corresponding primary quantities.

The Δ connection provides no neutral connection and each transformer must withstand full line-to-line voltage. The Δ connection does, however, provide a path for third-harmonic currents to flow. From Equations 4.36, we see that the third-harmonic currents in each transformer are all in phase with each

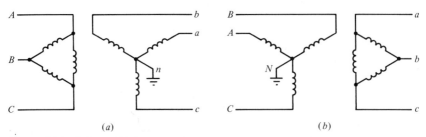

Figure 4.27 Δ-Y and Y-Δ connection.

other. By applying Kirchhoff's current law at each node of the transformer bank, we see that the third-harmonic currents circulate in the Δ connection and do not flow into the system. This phenomenon is a very desirable characteristic of Δ-connected transformers. Quite often Δ-connected tertiary windings are used in a Y-Y-connected transformer bank. This extra winding provides a path for the third-harmonic currents to circulate. This circulation prevents them from propagating into the system.

4.3.3 Δ-Y and Y-Δ Connections

The remaining three-phase transformer connections are Δ-Y and Y-Δ as shown in Figure 4.27. In both parts of Figure 4.27, the secondary windings are coupled with the primary windings drawn in parallel with them. In both of these connections each transformer still carries one-third of the power for a balanced load. However, the voltages being transformed are a mixture of line-to-line and line-to-neutral quantities. The currents are also a mixture of Δ phase currents and line currents. These combinations result in voltages and currents on different sides of the transformer bank being out of phase with each other. For instance, if a transformer bank is Δ-connected on the primary and Y-connected on the secondary, as it is in Figure 4.27a, the voltage V_{AB} is transformed to the voltage V_{an}. If the transformer is assumed to be ideal, then V_{AB} and V_{an} will be in phase. Under balanced conditions V_{ab} will be 30° out of phase with V_{an}. Therefore, V_{ab} will be 30° out of phase with V_{AB}. This is illustrated in Figure 4.28.

EXAMPLE 4.7

Three identical transformers, each with a ratio of transformation of 10, are to be connected to form a three-phase transformer bank. The high-voltage primary side has a balanced voltage of 1732 V. The balanced line

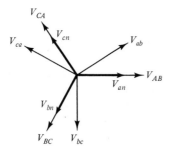

Figure 4.28 30° phase shift of Δ-Y connection.

4.3 THREE-PHASE TRANSFORMER BANKS

current on the primary side is 17.3 A. Assume ideal transformers, a power factor of 0.94 lag, and ABC phase sequence. With the primary voltage V_{AB} as reference, find the secondary voltage V_{ab} and secondary line current I_a if the transformers are connected in Y-Y, Δ-Δ, Y-Δ, and Δ-Y. Refer to Figures 4.25, 4.26, and 4.27 for transformer connections.

For a balanced ABC sequence, the primary phase voltage and phase current for phase a of a Y connection are

$$V_{AN} = \frac{V_{AB}}{\sqrt{3}} \underline{/-30°} = 1000\underline{/-30°} \text{ V}$$

$$I_A = 17.3\underline{/-30° - \cos^{-1}(0.94)} = 17.3\underline{/-50°} \text{ A}$$

The primary phase voltage and phase current for one phase of a Δ connection are

$$V_{AB} = 1732\underline{/0°} \text{ V}$$

$$I_{AB} = \frac{I_A}{\sqrt{3}}\underline{/30°} = 10\underline{/-20°} \text{ A}$$

For the Y-Y connection

$$V_{an} = \frac{V_{AN}}{10} = 100\underline{/-30°} \text{ V}$$

$$V_{ab} = \sqrt{3}V_{an}\underline{/30°} = 173.2\underline{/0°} \text{ V}$$

$$I_a = (10)I_A = 173\underline{/-50°} \text{ A}$$

For the Δ-Δ connection

$$V_{ab} = \frac{V_{AB}}{10} = 173.2\underline{/0°} \text{ V}$$

$$I_{ab} = (10)I_{AB} = 100\underline{/-20°} \text{ A}$$

$$I_a = \sqrt{3}I_{ab}\underline{/-30°} = 173\underline{/-50°} \text{ A}$$

For the Y-Δ connection

$$V_{ab} = \frac{V_{AN}}{10} = 100\underline{/-30°} \text{ V}$$

$$I_{ab} = (10)I_A = 173\underline{/-50°} \text{ A}$$

$$I_a = \sqrt{3}I_{ab}\underline{/-30°} = 300\underline{/-80°} \text{ A}$$

For the Δ-Y connection

$$V_{an} = \frac{V_{AB}}{10} = 173.2\underline{/0°} \text{ V}$$

$$V_{ab} = \sqrt{3}V_{an}\underline{/30°} = 300\underline{/30°} \text{ V}$$

$$I_a = (10)I_{AB} = 100\underline{/-20°} \text{ A}$$

■■

Connections in Δ-Y and Y-Δ pose particular problems in power system analysis problems and will be discussed in more detail in Chapter 6.

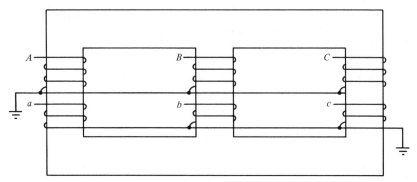

Figure 4.29 Core-type three-phase transformer.

4.3.4 Three-Phase Transformers

A three-phase transformer bank may consist of one three-phase transformer instead of three single-phase transformers. In a three-phase transformer all six windings are on a single core in one tank. A three-phase transformer is usually either a core or shell type. These two types of three-phase transformers are illustrated in Figures 4.29 and 4.30, respectively. Under steady-state conditions

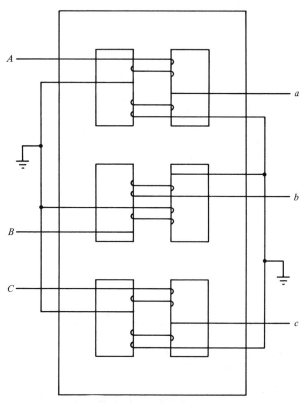

Figure 4.30 Shell-type three-phase transformer.

both of these types operate in the same manner as three single-phase transformers. Under faulted conditions the differences in available flux paths causes differences in operations of the two types. A three-phase transformer can be connected in any of the connections previously described for three single-phase transformers. Three-phase transformer ratings are always in line-to-line voltages and three-phase VA. The advantages of a three-phase transformer include lower cost, less space requirements, and less weight. Some flexibility is lost, however, in losing the ability to replace one single transformer should there be a failure in one phase only.

The incorporation of a three-phase transformer into system analysis calculations is done in the same way three single-phase transformer banks are handled. They are converted to equivalent Y connections, if necessary, and single-phase studies are performed. The other two phase quantities are simply shifted 120°. This process will be discussed in more detail in Chapter 6.

4.4 THREE-PHASE TRANSFORMATION WITH TWO TRANSFORMERS

If one transformer is removed from a Δ-Δ connection, the remaining configuration is called an open Δ. This connection is illustrated in Figure 4.31. Assume that the primary voltages in Figure 4.31 are $V_{AB} = V_1 \underline{/0°}$, $V_{BC} = V_1 \underline{/-120°}$, and $V_{CA} = V_1 \underline{/120°}$. The voltages transformed across the ideal transformers are $V_{ab} = V_2 \underline{/0°}$ and $V_{bc} = V_2 \underline{/-120°}$, where V_2 equals V_1 divided by the ratio of transformation. By Kirchhoff's voltage law, V_{ca} must equal $V_{cb} + V_{ba}$:

$$\begin{aligned} V_{ca} &= -(V_{ab} + V_{bc}) \\ &= -(V_2\underline{/0°} + V_2\underline{/-120°}) \\ &= V_2\underline{/120°} \end{aligned} \quad (4.37)$$

Therefore, the voltage V_{ca} for ideal transformers will be the same value whether or not the third transformer is in the bank. Under actual conditions a small unbalance in the voltages is introduced as the bank is loaded.

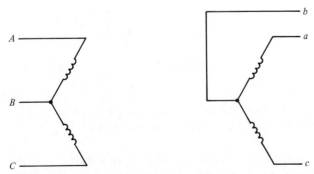

Figure 4.31 Open-Δ–open-Δ connection.

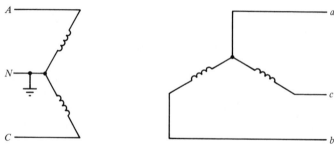

Figure 4.32 Open-Y–open-Δ connection.

If the maximum allowable current flow in each of three single-phase transformers composing a Δ-Δ connection is I, then the maximum line current is $\sqrt{3}I$. From Figure 4.31 we see that the maximum line current for an open Δ is restricted to the maximum allowable current flow in each transformer, I. Since the line-to-line voltages for both connections are the same, the kVA output of the open Δ is $1/\sqrt{3}$ or 58 percent of the kVA output of the full Δ bank.

In Figure 4.31 we see that only two voltages, which are 120° out of phase, are transformed from the primary to the secondary. Therefore, only two of the three primary phases would be required if a neutral is available. This open-Y–open-Δ connection is shown in Figure 4.32.

4.5 SUMMARY

In this chapter the basic operating principles of the power transformer were examined. Analysis of the operation of the transformer allowed development of models of varying complexity. The possible three-phase connections and their effect on voltage magnitude and phase were discussed. Since transformers can have a significant effect on power flow in a power system, they will be studied further in Chapter 6.

4.6 PROBLEMS

4.1. The curve in Figure 4.33 shows the voltage induced on a winding as a function of time. Draw the curve showing the flux linkages of the winding as a function of time when $\lambda(0) = 0$.

4.2. A 4000/230-V, 60-Hz, single-phase transformer is supplying a 11.5-kVA load at 230 V. If exciting current is neglected, determine the magnitude of the primary and secondary currents.

4.3. A 10-kVA, 480/120-V, single-phase transformer is subjected to a short circuit test and an open circuit test while being supplied from the high-voltage side. The results

4.6 PROBLEMS

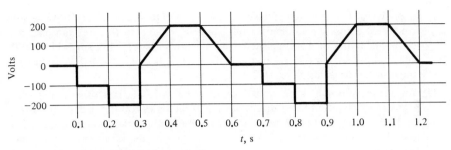

Figure 4.33 Voltage induced on winding.

are as follows:

Short Circuit Test	Open Circuit Test
$V_1 = 66.41$ V	$V_1 = 480$ V
$I_1 = 21$ A	$I_1 = 0.3$ A
$P_{in} = 441$ W	$P_{in} = 23$ W

If the dc resistance of the high-voltage coil is 16 times greater than the dc resistance of the low-voltage coil, draw the exact equivalent circuit of the transformer.

4.4. A 150-kVA, 2300/230-V, 60-Hz, single-phase transformer has the following constants:

$r_1 = 0.22\ \Omega$ $r_c = 6000\ \Omega$ $r_2 = 0.002\ \Omega$
$X_1 = 2.0\ \Omega$ $X_m = 1000\ \Omega$ $X_2 = 0.02\ \Omega$

Using the exact equivalent with all quantities referred to the high-voltage side, determine the difference between the magnitude of the secondary voltage at no load and the magnitude when it is supplying rated load at a power factor of 0.8 lag at a rated voltage of 230 V. This difference divided by the full-load voltage is called the voltage regulation of the transformer for that particular load.

4.5. The voltage on the secondary of a 5-kVA, 6900/230-V, 60-Hz, single-phase transformer increases by 8 percent of its full-load value of 230 V when its load is reduced from rated load at a power factor of 0.707 lag to no load. When the high-voltage winding is short-circuited, a voltage is impressed on the low-voltage winding of such magnitude that rated current flows in the low-voltage winding. Under these conditions the power input is 50 W. What is X_{eq} of the transformer? Use the approximate equivalent circuit.

4.6. A 60-Hz, single-phase induction motor draws 12.5 A at a voltage of 229 V and a power factor of 0.707 lag when supplied from the low-voltage side of a 3-kVA, 2400/240-V, 60-Hz, single-phase transformer. The transformer has a resistance of $r_{eq} = 40\ \Omega$ and a reactance of $X_{eq} = 150\ \Omega$. The transformer receives power over a rural line which has a resistance of 30 Ω/wire and a reactance of 95 Ω/wire. Using the series impedance model determine the magnitude of the voltage at the sending end of the line. (The single-phase transmission line has a supply and a return conductor.)

4.7. Four separate coils are wound around a common iron core. Three of the coils are identical. The other coil (coil 1) is rated at a higher voltage. The rating between the high-voltage coil and each of the other coils is 440/110 V. Coil 2 has a 10-Ω resistor connected across its terminals. Coil 3 has an inductor with a reactance of 5 Ω connected to it. Coil 4 is connected to a capacitor with a reactance of −11 Ω. With 440 V impressed on coil 1, determine the current flowing in coil 1. Assume that the transformer is ideal.

4.8. An autotransformer is connected so that the ratio of transformation is 1.25. Determine the ratio of the current flowing in the series coil to the current flowing in the common winding. Repeat the calculation for turns ratios of 2.0 and 2.5. The use of an autotransformer is most advantageous as the turns ratio approaches what value?

4.9. Four identical coils, each rated at 100 V, are connected in series on a common iron core. With each coil carrying 10 A, the total copper loss is found to be 40 W. With the autotransformer connected as shown in Figure 4.34, determine the total copper losses in the transformer when the resistor and capacitor each carry 8.0 A.

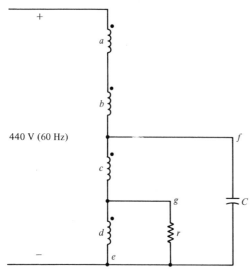

Figure 4.34 Autotransformer.

4.10. Three identical single-phase transformers, each having a rating of 200/40 kV, are connected in Y-Y and are supplied by a balanced three-phase source having a line-to-line voltage of 345 kV. The transformer bank is supplying a 60-MVA load at a power factor of 0.9 lag. Choose V_{AB} of the abc sequence source as reference. (a) Determine the magnitude and phase angles of all the line-to-line voltages, line currents, and currents flowing in the windings of the transformers. The transformers can be assumed to be ideal transformers. (b) Repeat part a when the secondaries of the transformers are connected in Δ.

4.11. The transformers of Problem 4.10 are connected in Δ-Δ and are supplied by a balanced, three-phase, acb sequence source with a line-to-line voltage of 180 kV. The transformer bank is supplying a 60-MVA load at a power factor of 0.9 lag. Choose V_{AB} of the source as reference. (a) Determine the magnitude and phase angles of all the line-to-line voltages, line currents, and currents flowing in the

4.6 PROBLEMS

windings of the transformers. The transformers can be assumed to be ideal. (b) Repeat part a for the secondaries of the transformers connected in Y.

4.12. Two identical transformers are connected in open-Y–open-Δ. Each transformer is rated at 790/440 V. They are supplying a balanced three-phase load of 12 kVA at a power factor of 0.85 lag and at a line-to-line voltage of 420 V. Determine the magnitude of the current flowing in each of the two phases and in the neutral of the high-voltage side. Consider the transformers to be ideal.

4.13. For the transformer bank shown in Figure 4.35, find the current in each transformer winding when

$$I_{SA} = 30\underline{/0°} \text{ A} \qquad V_{SA} = 100\underline{/0°} \text{ V}$$
$$I_{SB} = 0 \qquad V_{SB} = 100\underline{/-120°} \text{ V}$$
$$I_{SC} = 0 \qquad V_{SC} = 100\underline{/120°} \text{ V}$$

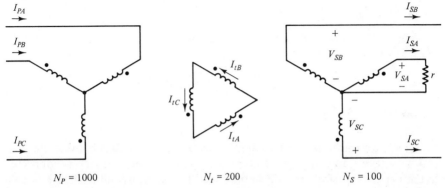

Figure 4.35 Transformer bank.

4.14. Three single-phase transformers have nameplate data of 138/13.8 kV, 10 MVA, and $X_{eq} = 190 \, \Omega$. They are connected Δ-Y with the Δ being the high-voltage side. Two balanced loads are connected to the low-voltage side. Load 1 is Δ connected and draws a total of 500 kVA at a power factor of 1.0. Load 2 is Y connected and draws 3 MVA in each phase at a power factor of 0.95 lag. Determine the magnitude of the line-to-line voltage on the high-voltage side of the transformer if the low-voltage side is maintained at a line-to-line value of 20 kV. Use the series reactance equivalent circuit.

4.15. A 10-kVA, 440/120-V, single-phase, two-winding transformer has a copper loss of 150 W, a hysteresis loss of 120 W, and an eddy current loss of 30 W when operating at full-load unity power factor with rated voltage. It is reconnected as a step-up autotransformer using the 400-V winding as the common coil. Determine the efficiency when the autotransformer is supplied from a 440-V source and is loaded with the largest unity power factor load that can be carried without the current in any coil exceeding rated value.

Chapter 5

Transmission Lines

5.1 INTRODUCTION

Power system transmission lines are the electrical connections between the generating stations and the load centers. The transmission network is connected to the step-up transformers within the generating stations and to the step-down transformers at the load stations. Figure 5.1 shows the relative position of transmission lines in a power system. The transmission lines that are energized to voltages of 345 kV and over are collectively referred to as the extra-high-voltage or EHV transmission system.

Figures 5.2 to 5.6 show several EHV lines. The line in Figure 5.2 is energized to 765 kV and is a bundled subconductor design. That is, each of the three phases consists of four separate subconductors in a cross-section configuration as shown in Figure 5.3. This kind of configuration serves to increase the effective radius of each phase in order to reduce high-voltage corona around the conductors. Figure 5.4 shows a closer view of a support tower for the line in which the subconductor bundling is clearly visible.

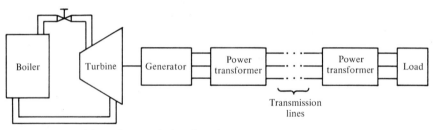

Figure 5.1 Position of transmission lines in power systems.

Figure 5.2 765-kV transmission line.

Figure 5.3 Four-subconductor configuration.

Figure 5.4 765-kV subconductors.

Figure 5.5 345-kV double-circuit transmission line.

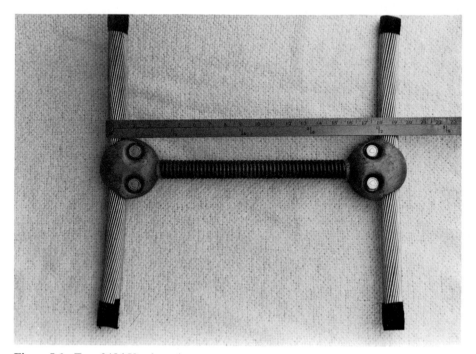

Figure 5.6 Two 345-kV subconductors.

5.1 INTRODUCTION

Figure 5.5 shows two separate transmission lines suspended from the same tower. These lines are energized to 345 kV and are also bundled subconductor designs. In this case each phase consists of two separate subconductors. Figure 5.6 shows a short section of this kind of bundling.

Another part of the transmission network consists of lines connecting the EHV system, through step-down transformers, to the distribution system transformers. This part is called the subtransmission network and is usually energized with voltages ranging from 34.5 kV to below 345 kV.

The distribution system is the part that connects the subtransmission system to the individual customer loads. Voltages in this system are usually 34.5 kV and lower. Figure 5.7 shows several distribution transmission lines attached to the same support pole. The line on the middle crossarm is tapped into an underground distribution line leading off the lower crossarm down the pole. The devices attached to the lower crossarm are voltage surge arresters and short circuit fuses protecting the distribution transformer and metering equipment.

All transmission lines in a power system exhibit the electrical properties of resistance, inductance, capacitance, and conductance. These properties have a major impact upon the operation of a system, and quantifying this impact is an important part of power system analysis. This chapter investigates the properties of resistance, inductance, and capacitance of transmission lines. Conductance is not considered here because it appears on a power system as

Figure 5.7 Distribution line.

leakage currents flowing across the transmission line insulators. These currents are negligible compared to the current flowing in the transmission lines and may be neglected.

5.2 TRANSMISSION LINE RESISTANCE

Series resistance of a transmission line is affected by the resistivity of its conductors, spiraling of the strands within its conductors, temperature, and skin effect. The resistivity of the material composing a conductor determines the dc resistance of that conductor:

$$r_{dc} = \frac{\rho l}{A} \quad \Omega \tag{5.1}$$

where r_{dc} = dc resistance
 ρ = resistivity, $\Omega \cdot m$
 l = conductor length, m
 A = cross-sectional area, m^2

Spiraling effects on resistance are explained by considering Figure 5.8, which shows a conductor cross section. The cross section shows that a conductor consists of individual strands of solid wire. Each layer of strands in Figure 5.8 is spiraled in the opposite direction of its adjacent layers. This spiraling holds the strands in place. However, because the strands must be longer than the actual length of the line in order to spiral them, the resistance of the transmission line is greater than the resistance due to the actual length of the line. The extra length increases resistance by 1 to 2 percent.

Resistance increases as temperature increases. The functional relationship between resistance and temperature is nearly linear over the range of temperature that a conductor is likely to experience. For aluminum conductors that relationship is given in Equation 5.2 for resistances r_1 and r_2 at temperature t_1 and t_2, respectively:

$$\frac{r_2}{r_1} = \frac{228 + t_2}{228 + t_1} \tag{5.2}$$

where t is the temperature in degrees centigrade.

The skin effect phenomenon causes an uneven distribution of current through a conductor cross section. If the current in a conductor were uniformly distributed through a conductor cross section, the resultant lines of flux would

Figure 5.8 Conductor cross section.

5.3 TRANSMISSION LINE INDUCTANCE

be concentric circles centered around the cross-sectional center of the conductor. Therefore, an incremental amount of current at the center of the conductor would have more flux linkages than the same amount of current near the outer edge of the conductor. Since flux linkage equals inductance multiplied by current, the inductance associated with the center of the conductor is larger than the inductance near the edge. Therefore, the ac current flow in a conductor tends to concentrate along its outer edge or "skin." The effective cross-sectional area of the conductor is reduced, and its effective resistance at 60 Hz is a few percent higher than its dc value.

5.3 TRANSMISSION LINE INDUCTANCE

Series inductance of a transmission line can be defined in terms of the flux linkage of its conductors:

$$\lambda_1 = L_1 I_1 + \sum_{j=2}^{N_c} M_{1j} I_j \tag{5.3}$$

where λ_1 = flux linkage of a conductor
L_1 = self-inductance of a conductor
I_1 = current flow in the conductor
M_{1j} = mutual inductance between the conductor and other nearby conductors
I_j = current flow in other conductors
N_c = number of conductors

Equation 5.3 shows that inductance of a transmission line consists of two components: self-inductance and mutual inductance. The self-inductance also has two components: internal inductance and external inductance. These quantities will be studied and specified in the following sections.

5.3.1 Internal Inductance

Each conductor of a transmission line has an internal inductance due to its own current flow. Figure 5.9 shows a conductor with an internal flux path at an arbitrary radius of x. Calculation of the internal flux linkage begins with the

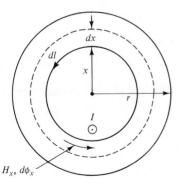

Figure 5.9   Internal inductance.

application of Ampere's law to the closed contour at radius x:

$$H_x 2\pi x = \frac{\pi x^2}{\pi r^2} I \tag{5.4}$$

where H_x = magnetic field intensity on the contour at radius x
$2\pi x$ = contour length at radius x
I = total conductor current flow
$\pi x^2/\pi r^2$ = fraction of the total conductor cross-section area contained within radius x
r = radius of the conductor

The result of using Ampere's law shows that the portion of the conductor current enclosed by contour x is found by assuming that I is distributed uniformly throughout the conductor cross section. This assumption ignores the skin effect on current distribution. However, at 60 Hz the error introduced by ignoring skin effect is small and considered negligible.

Solving Equation 5.4 for H_x yields

$$H_x = \frac{x}{2\pi r^2} I \tag{5.5}$$

The amount of flux associated with H_x can be found for a small region of thickness dx and a unit length of 1 m, as shown in Figure 5.9:

$$d\phi_x = \frac{\mu x I}{2\pi r^2} dx \tag{5.6}$$

The flux $d\phi_x$ links only the fraction of the conductor from the center to radius x. Therefore the flux linkage $d\lambda_x$ at radius x must account for this fraction by using the ratio $\pi x^2/\pi r^2$:

$$d\lambda_x = \frac{\mu x I}{2\pi r^2} dx \frac{\pi x^2}{\pi r^2}$$

$$= \frac{\mu I x^3}{2\pi r^4} dx \tag{5.7}$$

The total internal flux linkage is found by integrating $d\lambda_x$ from 0 to r:

$$\lambda_{int} = \int_0^r \frac{\mu I x^3}{2\pi r^4} dx$$

$$= \frac{\mu I}{8\pi} \quad \text{Wbt/m} \tag{5.8}$$

where λ_{int} is the internal flux linkage.

Conductors presently in use are made of metals that have permeabilities nearly equal to μ_0, so Equation 5.8 reduces to

$$\lambda_{int} = \frac{4\pi \times 10^{-7} I}{8\pi}$$

$$= \tfrac{1}{2} \times 10^{-7} I \quad \text{Wbt/m} \tag{5.9}$$

5.3 TRANSMISSION LINE INDUCTANCE

Internal inductance is calculated by recalling that

$$L_{int} = \frac{\lambda_{int}}{I}$$
$$= \tfrac{1}{2} \times 10^{-7} \text{ H/m} \qquad (5.10)$$

where L_{int} is the internal conductor inductance due only to the current flow in the conductor per unit length. Note that L_{int} is independent of the conductor radius r.

5.3.2 Inductance Due to External Flux Linkage

Each conductor of a transmission line has inductance due to external magnetic flux linkage. Figure 5.10 shows a conductor with an external flux path at an arbitrary radial distance x. Applying Ampere's law to the configuration of Figure 5.10 yields

$$H_x 2\pi x = I \qquad (5.11)$$

Flux density can be found from

$$B_x = \mu H_x$$
$$= \frac{\mu I}{2\pi x} \quad \text{Wb/m}^2 \qquad (5.12)$$

The flux passing through a thin section of area dx by 1 m can be calculated from B_x:

$$d\phi_x = \frac{\mu I}{2\pi x} dx \qquad (5.13)$$

Equation 5.13 can be used to solve for the flux between two radii D_1 and D_2 as shown in Figure 5.11.

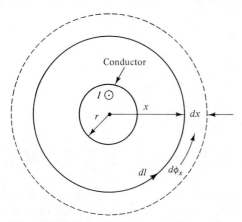

Figure 5.10 External flux path of a conductor.

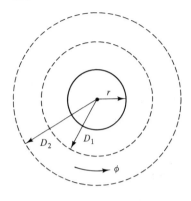

Figure 5.11 Flux linkage between D_1 and D_2.

$$\phi_{12} = \int_{D_1}^{D_2} \frac{\mu I}{2\pi x} dx$$
$$= \frac{\mu I}{2\pi} \ln \frac{D_2}{D_1} \quad \text{Wb/m} \quad (5.14)$$

The flux linkage follows as

$$\lambda_{12} = N\phi_{12}$$
$$= (1)(\phi_{12})$$
$$= \frac{\mu I}{2\pi} \ln \frac{D_2}{D_1} \quad \text{Wbt/m} \quad (5.15)$$

Inductance can be calculated from Equation 5.15 by dividing by current I:

$$L_{12} = \frac{\mu}{2\pi} \ln \frac{D_2}{D_1} \quad \text{H/m} \quad (5.16)$$

If the conductor is suspended in air, μ becomes μ_0.

$$L_{12} = 2 \times 10^{-7} \ln \frac{D_2}{D_1} \quad \text{H/m} \quad (5.17)$$

The total inductance of a conductor at an arbitrary distance D is

$$L_T = L_{\text{int}} + L_{\text{ext}}$$
$$= \tfrac{1}{2} \times 10^{-7} + 2 \times 10^{-7} \ln \frac{D}{r}$$
$$= 2 \times 10^{-7} \left(\ln e^{1/4} + \ln \frac{D}{r} \right)$$
$$= 2 \times 10^{-7} \ln \frac{D}{r'} \quad \text{H/m} \quad (5.18)$$

where r is the radius of the conductor and $r' = re^{-1/4}$; r' is a ficticious radius that accounts for internal inductance. A conductor made of the same material as the actual conductor but with a radius of r' instead of r would have an external inductance equal to the total inductance of the actual conductor of radius r.

5.3.3 Inductance of Single-Phase Transmission Lines

Inductance calculations shown to this point consider current flow in only one conductor. Power transmission is usually performed using two or more conductors for single-phase or multiphase systems. The close proximity of current flow in other conductors has an impact upon the flux linking each individual conductor. Figure 5.12 shows this situation for a single-phase transmission line using two conductors. The two conductors are each carrying I amperes but in opposite directions. They are separated by a distance D and have radii of r_1 and r_2 as shown. Calculation of flux linkage at a radial distance of x for the conductor on the left yields

$$\lambda_{\text{int}} = \tfrac{1}{2} I \times 10^{-7} \quad \text{Wbt/m} \quad 0 < x \leqslant r_1 \tag{5.19}$$

$$\lambda_{\text{ext}} = 2 \times 10^{-7} I \ln \frac{x}{r_1} \quad \text{Wbt/m} \quad r_1 < x \leqslant D - r_2 \tag{5.20}$$

These equations are the internal and external flux linkages derived in Equations 5.9 and 5.15. For distances greater than $D + r_2$, any flux path encloses a net current of 0. This condition means that the flux beyond $D + r_2$ does not link the circuit and cannot induce a voltage in it.

Equations 5.19 and 5.20 do not include the flux linkage calculation for the radial distance $D - r_2 \leqslant X \leqslant D + r_2$. Application of Ampere's law to this distance would yield a net current linkage on its right-hand side of I to 0 depending upon how much of the right conductor is enclosed in the contour. Since D is usually much greater than r_1 and r_2, a simplification of flux linkage calculation is made by assuming that current linked between the radial distances of r_1 to D is I. Therefore, the flux linkage for the conductor on the left is given by

$$\lambda_1 = \tfrac{1}{2} \times 10^{-7} I + 2 \times 10^{-7} I \ln \frac{D}{r_1}$$

$$= 2 \times 10^{-7} I \ln \frac{D}{r_1'} \quad \text{Wbt/m} \tag{5.21}$$

Its inductance is

$$L_1 = 2 \times 10^{-7} \ln \frac{D}{r_1'} \quad \text{H/m} \tag{5.22}$$

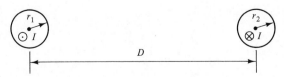

Figure 5.12 Single-phase transmission line.

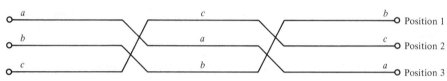

Figure 5.13 Transposition cycle.

Correspondingly, the inductance of the right conductor is

$$L_2 = 2 \times 10^{-7} \ln \frac{D}{r'_2} \quad \text{H/m} \tag{5.23}$$

5.3.4 Inductance of Three-Phase Transmission Lines

Calculations for three-phase transmission lines follow the same pattern as those for single-phase lines. However, an additional complication appears in three-phase lines as a result of transposition of the phases over the length of the line. Transposition is illustrated in Figure 5.13, which shows the three phases of a line as seen from above. The positions of phases a, b, and c change in a cycle that puts each phase in each of three possible positions over the span of the line. The reason for this cycle is to ensure that each phase experiences an equivalent amount of flux linkage. Thus, the average inductances of each of the phases will be equal, and a balanced voltage drop will occur along the line.

Not all three-phase transmission lines are transposed. Usually, shorter lines of lengths less than 80 km are not transposed because the differences in flux linkage of each phase is negligible over the shorter distances. Thus, assuming that these kinds of lines are transposed, as is done in this chapter, does not introduce large error in their analysis.

The calculation of inductance begins by solving for the flux linkage for one phase in each of its positions in the transposition cycle. Figure 5.14 shows positions 1, 2, and 3 and identifies distances between them. The calculation of flux linking phase a begins by identifying a point F that is remote from the line and used as a reference point, as shown in Figure 5.15. The flux linking phase a between a and F due only to I_a is

$$\lambda_{aFa} = 2 \times 10^{-7} I_a \ln \frac{D_{aF}}{r'_a} \quad \text{Wbt/m} \tag{5.24}$$

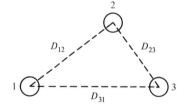

Figure 5.14 Positions of a transposition cycle.

5.3 TRANSMISSION LINE INDUCTANCE

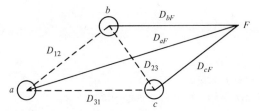

Figure 5.15 Reference point F.

where D_{aF} = distance between phase a and point F
  r'_a = effective radius of phase a conductor

The flux linking phase a out to point F due only to phase b current is

$$\lambda_{aFb} = 2 \times 10^{-7} I_b \ln \frac{D_{bF}}{D_{12}} \quad \text{Wbt/m} \tag{5.25}$$

where D_{bF} is the distance between phase b and F. The flux linking phase a out to point F due only to phase c current is

$$\lambda_{aFc} = 2 \times 10^{-7} I_c \ln \frac{D_{cF}}{D_{13}} \quad \text{Wbt/m} \tag{5.26}$$

where D_{cF} is the distance between phase c and F.

Adding λ_{aFa}, λ_{aFb}, and λ_{aFc} yields the total flux linking phase a out to point F.

$$\begin{aligned}\lambda_{aF} &= 2 \times 10^{-7} \left(I_a \ln \frac{D_{aF}}{r'_a} + I_b \ln \frac{D_{bF}}{D_{12}} + I_c \ln \frac{D_{cF}}{D_{13}} \right) \\ &= 2 \times 10^{-7} \left(I_a \ln \frac{1}{r'_a} + I_b \ln \frac{1}{D_{12}} + I_c \ln \frac{1}{D_{13}} \right. \\ &\quad \left. + I_a \ln D_{aF} + I_b \ln D_{bF} + I_c \ln D_{cF} \right) \end{aligned} \tag{5.27}$$

If point F is moved further away from the line, the three distances D_{aF}, D_{bF}, and D_{cF} become nearly equal:

$$D_{aF} \approx D_{bF} \approx D_{cF} = D_F \tag{5.28}$$

Equation 5.27 can now be reduced to

$$\lambda_{a1} = 2 \times 10^{-7} \left[I_a \ln \frac{1}{r'_a} + I_b \ln \frac{1}{D_{12}} + I_c \ln \frac{1}{D_{13}} + (I_a + I_b + I_c) \ln D_F \right] \tag{5.29}$$

If I_a, I_b, and I_c are balanced three-phase currents, their sum is 0 and Equation 5.29 reduces to

$$\lambda_{a1} = 2 \times 10^{-7} \left(I_a \ln \frac{1}{r'_a} + I_b \ln \frac{1}{D_{12}} + I_c \ln \frac{1}{D_{13}} \right) \quad \text{Wbt/m} \tag{5.30}$$

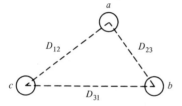

Figure 5.16 Second transposition cycle.

Figure 5.16 shows the next transposition cycle. Following the same development used for Equation 5.30, the flux linkage for phase a in the second transposition cycle is

$$\lambda_{a2} = 2 \times 10^{-7} \left(I_a \ln \frac{1}{r'_a} + I_b \ln \frac{1}{D_{23}} + I_c \ln \frac{1}{D_{21}} \right) \quad \text{Wbt/m} \quad (5.31)$$

The third transposition in the cycle is shown in Figure 5.17. Equation 5.32 shows the flux linkage of phase a in the third position.

$$\lambda_{a3} = 2 \times 10^{-7} \left(I_a \ln \frac{1}{r'_a} + I_b \ln \frac{1}{D_{31}} + I_c \ln \frac{1}{D_{32}} \right) \quad \text{Wbt/m} \quad (5.32)$$

The average flux linkage for phase a is

$$\lambda_a = \frac{\lambda_{a1} + \lambda_{a2} + \lambda_{a3}}{3}$$

$$= \tfrac{2}{3} \times 10^{-7} \left[3I_a \ln \frac{1}{r'_a} + (I_b + I_c) \ln \frac{1}{D_{12}D_{23}D_{31}} \right] \quad \text{Wbt/m} \quad (5.33)$$

Since the sum of I_a, I_b, and I_c is 0, the value of $I_b + I_c$ is $-I_a$. Using this relationship in Equation 5.33 reduces λ_a to

$$\lambda_a = \frac{2 \times 10^{-7}}{3} \left(3I_a \ln \frac{1}{r'_a} - I_a \ln \frac{1}{D_{12}D_{23}D_{31}} \right)$$

$$= 2 \times 10^{-7} I_a \ln \frac{\sqrt[3]{D_{12}D_{23}D_{31}}}{r'_a} \quad \text{Wbt/m} \quad (5.34)$$

The inductance of phase a follows as

$$L_a = 2 \times 10^{-7} \ln \frac{\sqrt[3]{D_{12}D_{23}D_{31}}}{r'_a} \quad \text{H/m} \quad (5.35)$$

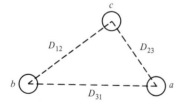

Figure 5.17 Third transposition cycle.

5.3 TRANSMISSION LINE INDUCTANCE

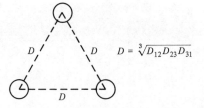

Figure 5.18 Symmetrically spaced conductors.

Calculations for L_b and L_c will yield nearly the same result:

$$L_b = 2 \times 10^{-7} \ln \frac{\sqrt[3]{D_{12}D_{23}D_{31}}}{r'_b} \quad \text{H/m} \tag{5.36}$$

$$L_c = 2 \times 10^{-7} \ln \frac{\sqrt[3]{D_{12}D_{23}D_{31}}}{r'_c} \quad \text{H/m} \tag{5.37}$$

If all three conductors have the same radius r, they will each have the same inductance L:

$$L = 2 \times 10^{-7} \ln \frac{\sqrt[3]{D_{12}D_{23}D_{31}}}{r'} \quad \text{H/m} \tag{5.38}$$

EXAMPLE 5.1

A transposed three-phase line has conductor spacing as shown. Calculate the inductance per phase:

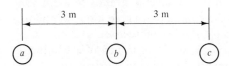

$$r_a = r_b = r_c = 0.01 \text{ m}$$
$$L = 2 \times 10^{-7} \ln \frac{\sqrt[3]{3 \times 3 \times 6}}{0.01 e^{-1/4}}$$
$$= 12.37 \times 10^{-7} \quad \text{H/m} \quad \blacksquare\blacksquare$$

The quantity $\sqrt[3]{D_{12}D_{23}D_{31}}$ is called the geometric mean distance or GMD. It is equivalent to the spacing of the conductors from each other if they were arranged in a symmetrical pattern as shown in Figure 5.18.

5.3.5 Geometric Mean Radius of Stranded Conductors

The conductor configurations considered so far have been for conductors with solid, round cross sections. As mentioned earlier in this chapter, most conductors consist of several strands intertwined as shown in Figure 5.8. Figure 5.19

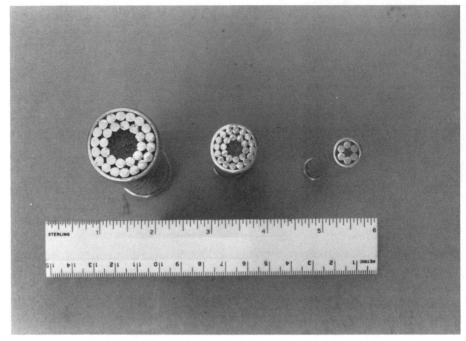

Figure 5.19 Stranded conductors.

shows a picture of a stranded conductor. This conductor consists of inner strands of steel (the darker strands) wrapped by several strands of aluminum. The inner steel strands add strength to the conductor. These kinds of conductors are referred to as aluminum conductor steel reinforced or ACSR.

The stranding can have an effect on the effective radius of a conductor. Since the conductor radius figures prominently in inductance calculations, some attention must be given to its value for stranded conductors. Consider a single-phase line consisting of two conductors X and Y, each with three strands arranged as shown in Figure 5.20. The strands are positioned so that they are in contact with their companions; they all have an equal radius r; and each strand is identical. Assuming that the current within each conductor divides equally among the strands, the inductance calculation for each conductor can be started in the same way as it is for three-phase lines. That is, treat each strand as if it is a separate conductor and calculate the flux linkage contributed by its own current and the current in the other conductors (strands).

The three strands of conductor X can be thought of as three equal inductances connected electrically in parallel. The total inductance of the con-

Figure 5.20 Single-phase line with stranded conductors.

5.4 CAPACITANCE OF TRANSMISSION LINES

ductor is

$$L_x = 2 \times 10^{-7} \ln \frac{\sqrt[9]{D_{x1y1}D_{x1y2}D_{x2y3}D_{x2y1}D_{x2y2}D_{x2y3}D_{x3y1}D_{x3y2}D_{x3y3}}}{\sqrt[9]{4(r')r^2 \times 4(r')r^2 \times 4(r')r^2}} \quad \text{Hm} \quad (5.39)$$

The quantity $\sqrt[9]{4^3(r')^3(r^2)^3}$ is referred to as the geometric mean radius or GMR of conductor X.

With a large number of strands the calculation of GMR can become very tedious. In addition, most conductors have steel strands or a metal alloy center for added strength. This results in an uneven current distribution. For these reasons the GMR of standard conductors are calculated and published in tables.

An additional comment is in order for the argument of the natural logarithm shown in Equation 5.39. The numerator of the argument is the GMD between conductors X and Y. All of the distances shown are between strands in conductor X and strands in conductor Y. If these distances are large relative to the radii of the strands, they can be approximated with negligible loss in accuracy by the distance between the geometric centers of conductors X and Y. If this distance is called D, the numerator of the argument of the natural logarithm reduces to

$$\sqrt[9]{(D^3)^3} = D \quad (5.40)$$

Similar reasoning can be applied to the inductance of three-phase transmission lines. Equation 5.38 becomes

$$L = 2 \times 10^{-7} \ln \frac{\sqrt[3]{D_{12}D_{23}D_{31}}}{r'_{eq}} \quad \text{H/m} \quad (5.41)$$

where r'_{eq} equals the GMR of the conductors in the line.

5.4 CAPACITANCE OF TRANSMISSION LINES

Transmission line conductors exhibit capacitance with respect to each other owing to the potential difference between them. The amount of capacitance between conductors is a function of conductor size, spacing, height above ground, and voltage. Capacitance of a conductor with respect to any other point is defined as

$$C = \frac{q}{v} \quad \text{F/m} \quad (5.42)$$

where q = instantaneous charge on the conductor, C/m
v = voltage drop from the conductor to a point of interest

Evaluation of Equation 5.42 proceeds by calculating v as a function of q and substituting it into the equation.

Potential difference v is defined as the line integral of the force in newtons acting on 1 C of positive charge as the charge is moved from one point to another. Force is exerted on the charge by the electric field between the two points. The electric field flux pattern surrounding an isolated cylindrical conductor is shown in Figure 5.21. The force acting on 1 C of charge in the field around the conductor is

$$F = qE = E \quad \text{N/m} \tag{5.43}$$

The electric field intensity for the conductor varies inversely with radial distance x and is expressed

$$E = \frac{q}{2\pi x \varepsilon} \quad \text{V/m} \tag{5.44}$$

where ε = the permittivity of the medium surrounding the conductor
q = charge on the conductor per meter

The potential difference between D_1 and D_2 shown in Figure 5.21 is calculated as

$$v_{12} = \int_{D_1}^{D_2} \frac{q}{2\pi x \varepsilon} \, dx = \frac{q}{2\pi \varepsilon} \ln \frac{D_2}{D_1} \quad \text{V} \tag{5.45}$$

where v_{12} is the voltage drop from D_1 to D_2.

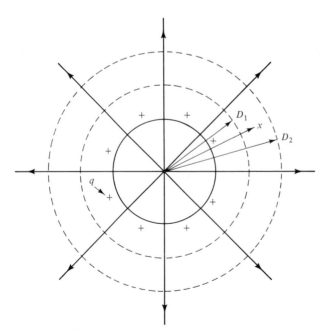

Figure 5.21 Electric field surrounding a charged cylindrical conductor.

5.4 CAPACITANCE OF TRANSMISSION LINES

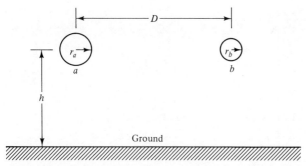

Figure 5.22 Single-phase line suspended aboveground.

5.4.1 Capacitance Calculations for Single-Phase Transmission Lines

Figure 5.22 shows a single-phase transmission line suspended h meters above ground. Calculation of capacitance between the conductors might at first glance seem to be a matter of calculating v_{ab} in terms of q_a as in Equation 5.45 and dividing the result into q_a. However, closer inspection of the electric field around conductors a and b reveals that it is distorted and does not have the uniform density shown in Figure 5.21 for an isolated conductor. Its distortion is caused by nonuniform charge density on conductors a and b. This distortion is illustrated in Figure 5.23. Some of the nonuniform charge distribution is caused by the close proximity of conductor a and b. However, calculations have shown that this distortion has a negligible effect on capacitance calculations and can be ignored. The most significant distortion is caused by the effect of the ground on the conductors.

The solution to this dilemma is to position "image conductors" beneath the ground surface in a symmetrical position with respect to conductors a and b. Figure 5.24 shows this position. The image conductors are assigned charges of opposite sign to that of their companions directly above them. They are also positioned at the same depth belowground as their companion conductors are aboveground. Figure 5.24 also shows the electric field between the four conductors without the ground being present. The electric field above the ground

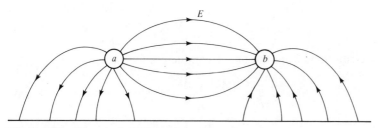

Figure 5.23 Electric field distortion.

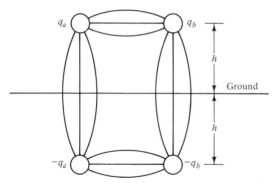

Figure 5.24 Image conductors.

looks exactly like the field shown in Figure 5.23. Therefore, capacitance calculations involving the four conductors of Figure 5.24 will yield accurate values without having to deal with ground surface distortion. This kind of modeling is called the method of images.

The calculation of capacitance between conductors begins by solving for voltage drop V_{ab}. This calculation is performed by moving 1 C of charge from conductor a to b and integrating the line integral of the force acting on the charge caused by the electric field between them. Equation 5.45 shows the result of this operation for a uniformly distributed field around a cylindrical conductor. This equation can be used directly for the system of conductors shown in Figure 5.24 if the contribution of V_{ab} from each conductor's field is calculated separately. Then these four components of V_{ab} can be added to yield the total V_{ab}. For example, setting the charge on conductor b, conductor b's image, and the image conductor of a to zero leaves only the electric field due to conductor a. Then, the voltage drop for this situation is

$$V_{ab}^{q_a} = \frac{q_a}{2\pi\varepsilon} \ln \frac{D}{r_a} \quad \text{V} \tag{5.46}$$

where $V_{ab}^{q_a}$ = voltage drop due to charge only on conductor a
r_a = radius of conductor a

For charge only on conductor b

$$V_{ab}^{q_b} = \frac{q_b}{2\pi\varepsilon} \ln \frac{r_b}{D} \quad \text{V} \tag{5.47}$$

The voltage contributions from the image conductors follow as

$$V_{ab}^{-q_a} = \frac{-q_a}{2\pi\varepsilon} \ln \frac{\sqrt{4h^2 + D^2}}{2h} \quad \text{V} \tag{5.48}$$

$$V_{ab}^{-q_b} = \frac{-q_b}{2\pi\varepsilon} \ln \frac{2h}{\sqrt{4h^2 + D^2}} \quad \text{V} \tag{5.49}$$

5.4 CAPACITANCE OF TRANSMISSION LINES

The total value of V_{ab} is the sum of Equations 5.46 to 5.49:

$$V_{ab} = \frac{q_a}{2\pi\varepsilon} \ln \frac{D}{r_a} + \frac{q_b}{2\pi\varepsilon} \ln \frac{r_b}{D} + \frac{-q_a}{2\pi\varepsilon} \ln \frac{\sqrt{4h^2 + D^2}}{2h}$$

$$+ \frac{-q_b}{2\pi\varepsilon} \ln \frac{2h}{\sqrt{4h^2 + D^2}} \qquad (5.50)$$

For a single-phase line $q_a = -q_b$ and Equation 5.50 reduces to

$$V_{ab} = \frac{q_a}{2\pi\varepsilon} \left(\ln \frac{D^2}{r_a r_b} - \ln \frac{4h^2 + D^2}{4h^2} \right) \quad \text{V} \qquad (5.51)$$

Capacitance C_{ab} follows from Equation 5.42:

$$C_{ab} = \frac{2\pi\varepsilon}{\ln \dfrac{D^2}{r_a r_b} - \ln \dfrac{4h^2 + D^2}{4h^2}} \quad \text{F/m} \qquad (5.52)$$

EXAMPLE 5.2
Calculate the line-to-line capacitance for the single-phase line shown:

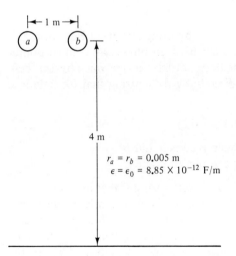

$r_a = r_b = 0.005$ m
$\varepsilon = \varepsilon_0 = 8.85 \times 10^{-12}$ F/m

Using Equation 5.52,

$$C_{ab} = \frac{(2\pi)8.85 \times 10^{-12}}{\ln(1^2/0.005^2) - \ln[4(4)^2 + 1^2/4(4)^2]}$$
$$= 5.255 \times 10^{-12} \text{ F/m} \qquad \blacksquare\blacksquare \quad (5.53)$$

Note the second term in the denominator of Equations 5.52 and 5.53. This term is the contribution of the ground effect to the capacitance and has a value of nearly zero. This situation will always occur for transmission lines that have conductor heights comparatively higher than conductor spacing. For these conditions the ratio

$$\frac{4h^2 + D^2}{4h^2} \approx 1.0$$

will prevail, and the natural logarithm of the ratio will be nearly zero. Under these conditions, which are standard for most transmission lines, the effect of the earth can be neglected.

5.4.2 Capacitance Calculations for Three-Phase Transmission Lines

Capacitance calculations for three-phase transmission lines are usually performed in a way to yield the line-to-neutral value for each conductor. This method requires solving for the line-to-neutral voltage of each conductor as a function of the conductor's charge and using Equation 5.54:

$$C_n = \frac{q_a}{V_{an}} \tag{5.54}$$

where V_{an} = line-to-neutral voltage of phase a
q_a = charge on phase a, a phasor quantity, C/m

Solving for V_{an} cannot be performed directly by using a form of Equation 5.45 because distances used in this equation are between physically defined points. The neutral point in a three-phase line does not have a rigorous physical definition. However, V_{an} can be found indirectly by recognizing that for balanced three-phase voltages

$$V_{an} = \tfrac{1}{3}(V_{ab} + V_{ac}) \tag{5.55}$$

where V_{ab} = line-to-line voltage between phases a and b
V_{ac} = line-to-line voltage between phases a and c

Solutions for V_{ab} and V_{ac} can be obtained directly using Equation 5.45.

Consider the calculation of V_{ab} for a three-phase transmission line with conductor spacing as shown in Figure 5.25. Assume that the line is transposed and that the height of the line makes ground effect negligible. Using the same method that was used for inductance calculations, the calculation of V_{ab} for the entire line begins by considering the value of V_{ab} along each section of the transposition cycle. Using the conductor positions shown in Figure 5.25 as the first

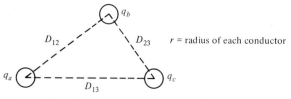

Figure 5.25 Three-phase transmission line.

5.4 CAPACITANCE OF TRANSMISSION LINES

portion of the cycle,

$$V_{ab}^1 = \frac{1}{2\pi\varepsilon}\left(q_a \ln \frac{D_{12}}{r} + q_b \ln \frac{r}{D_{21}} + q_c \ln \frac{D_{32}}{D_{31}}\right) \quad \text{V} \quad (5.56)$$

where V_{ab}^1 is the voltage in the first position.

Note that the terms enclosed by the parentheses are the separate voltage contributions due to charge only on phase a, b, and c, respectively. Figure 5.26a and b show the second and third portions of the transposition cycle, respectively. Equations 5.57 and 5.58 finish the V_{ab} component calculations for the second and third portions of the transposition cycle:

$$V_{ab}^2 = \frac{1}{2\pi\varepsilon}\left(q_a \ln \frac{D_{23}}{r} + q_b \ln \frac{r}{D_{32}} + q_c \ln \frac{D_{13}}{D_{12}}\right) \quad \text{V} \quad (5.57)$$

$$V_{ab}^3 = \frac{1}{2\pi\varepsilon}\left(q_a \ln \frac{D_{31}}{r} + q_b \ln \frac{r}{D_{13}} + q_c \ln \frac{D_{21}}{D_{23}}\right) \quad \text{V} \quad (5.58)$$

At this point an assumption concerning the charges shown in Equations 5.56 to 5.58 must be made. This assumption is that the charge on each conductor is the same in every portion of the transposition cycle. This assumption is made so that an average value of V_{ab} can be calculated using Equations 5.56 to 5.58 without creating unknown charges other than q_a, q_b, and q_c. However, using this assumption means that V_{ab}^1, V_{ab}^2, and V_{ab}^3 will not be equal, and if voltage drops along the conductors are neglected, this situation violates Kirchhoff's voltage law. This problem is not as serious as it seems because capacitance calculations completed under these conditions are reasonably accurate when compared to calculations performed without the charge equality assumption.

The average value of V_{ab} is found using Equations 5.56 to 5.58 and is shown in Equation 5.59:

$$V_{ab} = \frac{1}{(3)2\pi\varepsilon}\left(q_a \ln \frac{D_{12}D_{23}D_{31}}{r^3} + q_b \ln \frac{r^3}{D_{12}D_{23}D_{31}} + q_c \ln \frac{D_{12}D_{23}D_{31}}{D_{12}D_{23}D_{31}}\right) \quad (5.59)$$

Note that as in the inductance calculation, the GMD of the conductors appears in the natural logarithm arguments.

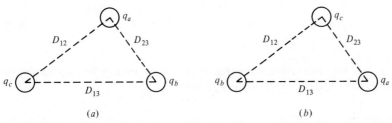

Figure 5.26 Remaining portions of transposition cycle. (a) Second portion of cycle. (b) Third portion of cycle.

The next step in the capacitance calculation is to solve for V_{ac}. Using the same charge equality assumption as for V_{ab}, V_{ac} appears as

$$V_{ac} = \frac{1}{(3)2\pi\varepsilon}\left(q_a \ln\frac{D_{12}D_{23}D_{31}}{r^3} + q_b \ln\frac{D_{12}D_{23}D_{31}}{D_{12}D_{23}D_{31}} + q_c \ln\frac{r^3}{D_{12}D_{23}D_{31}}\right) \quad (5.60)$$

Operating on V_{ab} and V_{ac} as shown in Equation 5.55 yields V_{an}:

$$V_{an} = \frac{1}{(9)2\pi\varepsilon}\left(2q_a \ln\frac{D_{12}D_{23}D_{31}}{r^3} + q_b \ln\frac{r^3}{D_{12}D_{23}D_{31}} + q_c \ln\frac{r^3}{D_{12}D_{23}D_{31}}\right) \quad (5.61)$$

In a balanced three-phase operation

$$q_a + q_b + q_c = 0 \qquad q_a = -q_b - q_c \quad (5.62)$$

and V_{an} reduces to

$$V_{an} = \frac{1}{2\pi\varepsilon}\left(q_a \ln\frac{\sqrt[3]{D_{12}D_{23}D_{31}}}{r}\right) \quad V \quad (5.63)$$

Then, the calculation of C_n follows from Equation 5.54:

$$C_n = \frac{2\pi\varepsilon}{\ln\sqrt[3]{D_{12}D_{23}D_{31}}/r} \quad F/m \quad (5.64)$$

EXAMPLE 5.3

Calculate the line-to-neutral capacitance of the transposed transmission line shown below:

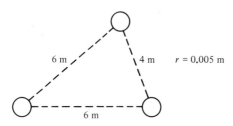

Using Equation 5.64,

$$C_n = \frac{(2\pi)8.85 \times 10^{-12}}{\ln[\sqrt[3]{6(4)6}/0.005]}$$
$$= 8 \times 10^{-12} \quad F/m \quad \blacksquare\blacksquare$$

EXAMPLE 5.4

Calculate the line-to-neutral capacitance of the three-phase transmission line shown, neglecting the effects of earth:

|←— 3 m —→|←— 3 m —→|

○   ○   ○ $r = 0.004$ m

Using Equation 5.64 and the data shown,

$$C_n = \frac{(2\pi)8.85 \times 10^{-12}}{\ln[\sqrt[3]{3(3)6}/0.004]}$$
$$= 8.116 \times 10^{-12} \quad \text{F/m}$$

■ ■

5.5 BUNDLED CONDUCTORS IN THREE-PHASE TRANSMISSION LINES

Extra-high-voltage transmission systems usually operate at or above 345 kV. For these higher ranges of voltages, the high-intensity electric fields around the conductors cause ionization of the air close to the conductor surfaces. This ionization is called corona. Power loss along the line is one of many bad features of corona. Additionally some of its more annoying and undesirable effects are radio communication interference and audible noise, which has been likened to "frying bacon" or "a swarm of angry bees."

Corona can be reduced by using bundled conductors to create electric field cancellation in the high-intensity area near the surface of conductors. Bundled conductors consist of two or more subconductors arranged in configurations like those shown in Figure 5.27.

Evaluation of inductance of bundled conductor lines is easily performed by using Equation 5.41 with some slight modification to r'_{eq}:

$$L = 2 \times 10^{-7} \ln \frac{\sqrt[3]{D_{12}D_{23}D_{31}}}{r^{B'}_{eq}} \tag{5.65}$$

where $r^{B'}_{eq}$ is the GMR of the subconductor bundle in each phase. Also, the distances between conductor positions can be approximated with negligible error as being the distances measured between the centers of the subconductor bundles.

The GMR for a set of n subconductors is the n^2 root of the quantity, the product of the distance from each subconductor to every other subconductor times the r' of every subconductor. Referring to Figure 5.27, if all the conductors are identical, then for the two-subconductor bundle

$$\text{GMR} = \sqrt[4]{ddr'r'} = \sqrt{dr'} \tag{5.66}$$

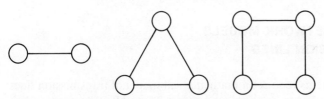

Figure 5.27 Examples of subconductor configurations.

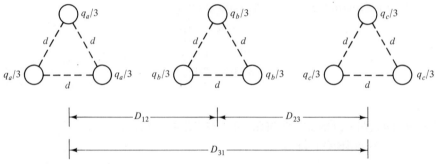

Figure 5.28 Transmission line with bundled conductors.

For the three-subconductor bundle

$$\text{GMR} = \sqrt[9]{dddddddr'r'r'} = \sqrt[3]{ddr'} \tag{5.67}$$

The GMR for the four-subconductor is, similarly,

$$\text{GMR} = \sqrt[16]{(r'\ ddd2^{1/2})^4} = 1.09 \sqrt[4]{d^3 r'} \tag{5.68}$$

The effects of bundled conductors in transmission lines can be included in capacitance calculations by assuming that the charge on each conductor divides evenly among its subconductors. This assumption is not precisely accurate because the influence of charge on other conductors will cause slight imbalances of charge distribution among companion subconductors. However, the inaccuracy of capacitance calculations made by assuming evenly distributed charge has been found to be negligible when compared to more rigorous calculations made without the assumption.

Figure 5.28 shows a three-phase transmission line with bundled conductors. The derivation for capacitance of bundled lines follows the same steps as the derivation for lines with one conductor per phase, with the exception that the expressions for V_{ab} and V_{ac} contain nine terms. Each term is the voltage contribution from charge on each of the nine subconductors. The result of the derivation is given in Equation 5.69:

$$C_n = \frac{2\pi\varepsilon}{\ln(\sqrt[3]{D_{12}D_{23}D_{31}}/r^B_{\text{eq}})} \quad \text{F/m} \tag{5.69}$$

where r^B_{eq} is the GMR of the bundle. This quantity is calculated as shown earlier for inductance with the exception that the radius r of each subconductor is used instead of r'.

5.6 ELECTRICAL NETWORK MODELS OF TRANSMISSION LINES

Analysis of power systems requires a mathematical model of transmission lines in order to calculate voltages, currents, and power flows. The transmission line models must account for the series resistance and inductance and shunt ca-

5.6 ELECTRICAL NETWORK MODELS OF TRANSMISSION LINES

pacitance of each phase. These quantities are distributed along the entire length of the line. Calculations of voltages and currents at any point on the line can be done with reference to Figure 5.29. At distance x meters from the receiving end, voltage drop dV is caused by current flow I through the series impedance:

$$z = r + j\omega L \quad (5.70)$$

where r = series resistance, Ω/m
ω = angular frequency of the voltages and currents, rad/s
L = series inductance, H/m

$$dV = Iz\,dx \quad (5.71)$$

or
$$\frac{dV}{dx} = Iz \quad (5.72)$$

Current dI flows through the shunt admittance y:

$$y = j\omega C \quad V/m$$

where C is the line-to-neutral or shunt capacitance in farads per meter. Then

$$dI = Vy\,dx \quad (5.73)$$

or
$$\frac{dI}{dx} = yV \quad (5.74)$$

Differentiating Equations 5.72 and 5.74 with respect to x results in

$$\frac{d^2V}{dx^2} = z\frac{dI}{dx} \quad (5.75)$$

$$\frac{d^2I}{dx^2} = y\frac{dV}{dx} \quad (5.76)$$

The values of dI/dx and dV/dx can be substituted into Equations 5.75 and 5.76:

$$\frac{d^2V}{dx^2} = yzV \quad (5.77)$$

$$\frac{d^2I}{dx^2} = yzI \quad (5.78)$$

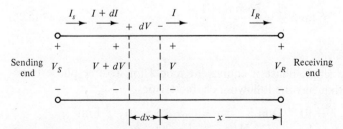

Figure 5.29 One phase of a transmission line.

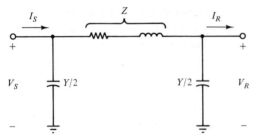

Figure 5.30 Equivalent π circuit of one phase of a transmission line.

Equations 5.77 and 5.78 can be solved for V and I with the boundary conditions

$$V|_{x=0} = V_R \qquad I|_{x=0} = I_R \qquad (5.79)$$

The result is

$$V(x) = \tfrac{1}{2}(V_R + I_R Z_c)e^{\gamma x} + \tfrac{1}{2}(V_R - I_R Z_c)e^{-\gamma x} \qquad (5.80)$$

$$I(x) = \tfrac{1}{2}(V_R/Z_c + I_R)e^{\gamma x} - \tfrac{1}{2}(V_R/Z_c - I_R)e^{-\gamma x} \qquad (5.81)$$

where $Z_c = \sqrt{z/y}$ = the characteristic impedance
$\gamma = \sqrt{zy}$ = the propagation constant

When only the sending end and receiving end quantities are of interest, these equations can be used to model each phase of a line as an equivalent π circuit, as shown in Figure 5.30. The values of Z and Y must be calculated such that when V_R and I_R are specified, V_S and I_S will be equal to their calculated values using Equations 5.80 and 5.81. For a transmission line that is l meters long, Z and Y are calculated as

$$Z = Z_c \sinh \gamma l \qquad (5.82)$$

$$Y = \frac{2}{Z_c} \tanh \frac{\gamma l}{2} \qquad (5.83)$$

Since γ is normally a complex number, the solution of the hyperbolic functions is not trivial. However, these functions with complex arguments can be expressed as functions with real arguments.

$$\sinh(\alpha l + j\beta l) = \sinh \alpha l \cos \beta l + j \cosh \alpha l \sin \beta l \qquad (5.84)$$

$$\cosh(\alpha l + j\beta l) = \cosh \alpha l \cos \beta l + j \sinh \alpha l \sin \beta l \qquad (5.85)$$

$$\tanh \frac{\gamma l}{2} = \frac{\cosh \gamma l - 1}{\sinh \gamma l} \qquad (5.86)$$

EXAMPLE 5.5

Calculate the single-phase π equivalent model for a three-phase transmission line that has the following characteristics:

$$r = 10^{-4} \ \Omega/m \qquad C = 8.5 \times 10^{-12} \ F/m$$
$$L = 1.4 \times 10^{-6} \ H/m \qquad l = 200 \ km$$

5.6 ELECTRICAL NETWORK MODELS OF TRANSMISSION LINES

First calculate z and y:

$$z = 10^{-4} + (2\pi 60)1.4 \times 10^{-6}$$
$$= (1 + j5.278) \times 10^{-4} \ \Omega/\text{m}$$

$$y = (2\pi 60)8.5 \times 10^{-12}$$
$$= j3.204 \times 10^{-9} \ \text{V}/\text{m}$$

Next calculate Z_c and γ:

$$Z_c = \sqrt{z/y}$$
$$= \sqrt{\frac{(1 + j5.278) \times 10^{-4}}{j3.204 \times 10^{-9}}}$$
$$= 407.7 - j38.28$$

$$\gamma = \sqrt{zy}$$
$$= \sqrt{(1 + j5.278) \times 10^{-4}(j3.204 \times 10^{-9})}$$
$$= (0.1232 + j1.306) \times 10^{-6}$$

Finally calculate Z and Y:

$$Z = Z_c \sinh \gamma l$$
$$= (407.7 - j38.28)\sinh[(0.1232 + j1.306) \times 10^{-6}(2 \times 10^5)]$$
$$= 19.58 + j104.4 \ \Omega$$

$$Y = \frac{2(\cosh \gamma l - 1)}{Z_c \sinh \gamma l}$$
$$= \frac{2\cosh[(0.1232 + j1.306) \times 10^{-6}(2 \times 10^5)] - 2}{(407.7 - j38.28)\sinh[(0.1232 + j1.306) \times 10^{-6}(2 \times 10^5)]}$$
$$= j6.439 \times 10^{-4} \ \text{V}$$

The equivalent π model now appears as

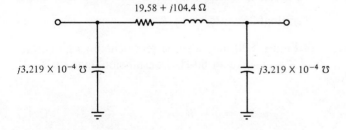

■ ■

An interesting point concerning the calculation of Z and Y for transmission lines is that for values of l in the range of 80 to 250 km, the Z and Y values can be approximated as

$$Z = zl \tag{5.87}$$

$$Y = yl \tag{5.88}$$

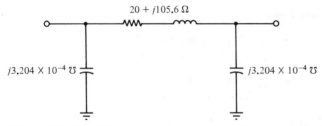

Figure 5.31 Equivalent π circuit using approximate values of Z and Y.

with reasonable accuracy. For example, the π model of the transmission line in Example 5.5 is shown in Figure 5.31 with Z and Y calculated from Equations 5.87 and 5.88.

For lines less than 80 km, the value of Z can still be calculated using Equation 5.87. However, Y is small enough when compared to Z that it can be neglected. For lines longer than 250 km, Equations 5.82 and 5.83 should be used.

5.7 REAL- AND REACTIVE-POWER FLOW ON A TRANSMISSION LINE

Given the terminal voltages magnitudes and angles and the impedances of the transmission line model of Figure 5.30, the flow of real and reactive power from one terminal toward the other terminal can be calculated. To illustrate this calculation and determine those variables which effect these flows the most, Figure 5.30 will be used as an example with the left terminal designated as terminal 1 and the right as terminal 2. The phase voltages at each terminal will be designated

$$\text{Terminal 1:} \quad V_{an} = V_1 \underline{/\theta_1} \tag{5.89}$$

$$\text{Terminal 2:} \quad V_{an} = V_2 \underline{/\theta_2} \tag{5.90}$$

The series impedance Z of Figure 5.30 may also be represented as an admittance Y_{12}. Since the angle of Z is positive for 60-Hz transmission lines, the angle of Y_{12} will be negative:

$$\frac{1}{Z} = Y_{12} = g - jb \tag{5.91}$$

The complex power flowing from terminal 1 toward terminal 2 is the summation of the flows in the shunt branch at terminal 1 and the series branch connecting terminals 1 and 2. The complex power flow in the shunt branch is

$$\begin{aligned} S_{S1} &= V_1 \underline{/\theta_1} [(Y/2)(V_1 \underline{/\theta_1})]^* \\ &= V_1^2 (Y/2)^* \end{aligned} \tag{5.92}$$

5.7 REAL- AND REACTIVE-POWER FLOW ON A TRANSMISSION LINE

Since conductance is neglected for power transmission lines, $Y/2$ is purely reactive and S_{S1} is composed of reactive power only.

The complex power flow in the series branch is

$$S_{12} = V_1 \underline{/\theta_1} [(V_1 \underline{/\theta_1} - V_2 \underline{/\theta_2}) Y_{12}]^* \qquad (5.93)$$

Equation 5.93 can be rewritten

$$S_{12}^* = V_1 \underline{/-\theta_1} (V_1 \underline{/\theta_1} - V_2 \underline{/\theta_2})(g - jb) \qquad (5.94)$$

After the appropriate multiplications,

$$S_{12}^* = [V_1^2 g - V_1 V_2 g \cos(\theta_2 - \theta_1) - V_1 V_2 b \sin(\theta_2 - \theta_1)] \\ + j[-V_1^2 b + V_1 V_2 b \cos(\theta_2 - \theta_1) - V_1 V_2 g \sin(\theta_2 - \theta_1)] \qquad (5.95)$$

Equation 5.95 can now be separated into its real and imaginary parts. Therefore, by changing the sign of the imaginary part to compensate for the conjugate, they become

$$P_{12} = V_1^2 g - V_1 V_2 g \cos(\theta_2 - \theta_1) - V_1 V_2 b \sin(\theta_2 - \theta_1) \qquad (5.96)$$

$$Q_{12} = V_1^2 b - V_1 V_2 b \cos(\theta_2 - \theta_1) + V_1 V_2 g \sin(\theta_2 - \theta_1) \qquad (5.97)$$

Now examine the terms of Equation 5.96 more closely. Under normal operating conditions V_1 and V_2 will be nearly equal in magnitude. The same condition is true for θ_1 and θ_2, where a difference of 15° would be considered large. Therefore, the following approximations of the trigonometrical functions can be made:

$$\cos(\theta_2 - \theta_1) \approx 1.0 \qquad (5.98)$$

$$\sin(\theta_2 - \theta_1) \approx (\theta_2 - \theta_1) \qquad (5.99)$$

Equation 5.96 can be rearranged to read

$$P_{12} = V_1(V_1 - V_2)g - V_1 V_2 b(\theta_2 - \theta_1) \qquad (5.100)$$

Recall that the value of series inductive reactance for a power transmission line is normally much greater than the series resistance. Therefore, b will be much greater than g. If the common V_1 is factored from both terms of Equation 5.100, the first term is left with two small numbers multiplied together $V_1 - V_2$ and g. The second term has one small number $\theta_2 - \theta_1$, but it is multiplied by two relatively large numbers, V_2 and b. Therefore, the second term appears to dominate the flow of real power from terminal 1 toward terminal 2:

$$P_{12} = V_1 V_2 b(\theta_1 - \theta_2) \qquad (5.101)$$

Using the same reasoning, the flow of reactive power from terminal 1 toward terminal 2 would be dominated by the term

$$Q_{12} = V_1 b(V_1 - V_2) \qquad (5.102)$$

Examination of Equation 5.101 shows that while small changes in V_1 or V_2 will not drastically change P_{12}, small changes in θ_1 or θ_2 will have a significant effect on real-power flow. By the same reasoning, small changes in V_1 or

V_2 will affect reactive-power flow, while small changes in θ_1 or θ_2 will have little influence. Remember from Equation 5.92 that the reactive-power flow in the shunt branch of the model is dependent only on the terminal voltage magnitude. Therefore, the following conclusion for normal operating conditions can be made:

1. The flow of real power on a transmission line is determined by the angle difference of the terminal voltages.
2. The flow of reactive power is determined by the magnitude difference of terminal voltages.

This concept will be discussed again in Chapter 6 along with control of complex power flow.

5.8 SUMMARY

In this chapter a method of calculating the variables that affect the resistance, inductance, and capacitance of a transmission line were presented. Mathematical models of transmission lines for use in computer programs were derived. Finally those quantities that affect real- and reactive-power flow on a transmission line were discussed. The terminal voltage magnitudes most directly affected the flow of reactive power, and the terminal voltage angles controlled real-power flow.

5.9 PROBLEMS

5.1. Calculate the inductance of the single-phase transmission line shown in Figure 5.32.

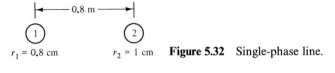

$r_1 = 0.8$ cm $\quad r_2 = 1$ cm **Figure 5.32** Single-phase line.

5.2. Calculate the inductance of the single-phase transmission line shown in Figure 5.33.

$\longleftarrow 0.75$ m $\longrightarrow$

① ②

$r_1 = 0.9$ cm $\quad r_2 = 0.9$ cm **Figure 5.33** Single-phase line.

5.3. Calculate the voltage induced on the telephone circuit from the single-phase line shown in Figure 5.34.

5.9 PROBLEMS

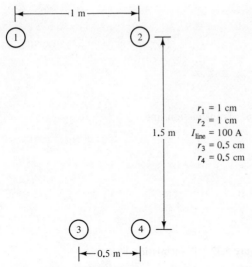

Figure 5.34 Single-phase line and telephone circuit.

5.4. Calculate the voltage induced on the telephone circuit from the single-phase line shown in Figure 5.35.

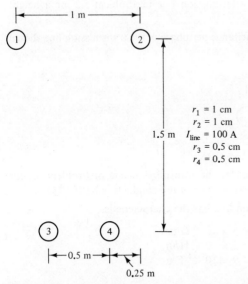

Figure 5.35 Single-phase line and telephone circuit.

5.5. Calculate the inductance per phase of the transposed transmission line with cross section as shown in Figure 5.36.

5.6. Calculate the inductance per phase of the transposed transmission line with cross section as shown in Figure 5.37.

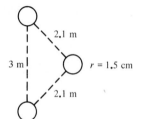

Figure 5.36 Transmission line.

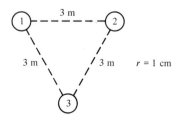

Figure 5.37 Transmission line.

5.7. Calculate the capacitance per phase of the transmission line in Problem 5.5. Neglect ground effects.

5.8. Calculate the capacitance per phase of the transmission line in Problem 5.6. Neglect ground effects.

5.9. Calculate the capacitance of the transmission line in Problem 5.2 for a height aboveground of 3 m.

5.10. Calculate the inductance and capacitance per phase of the transmission line shown in Figure 5.38.

Figure 5.38 Transmission line.

5.11. Determine the π equivalent model for the transmission line of Problem 5.5 for lengths of 50 and 160 km. Let the resistance of the conductors be 10^{-4} Ω/m.

5.12. A three-phase, 300-km transmission line has the characteristics

$$R = 5 \times 10^{-5} \text{ }\Omega/\text{m}$$
$$L = 2 \times 10^{-6} \text{ H/m}$$
$$C = 9 \times 10^{-12} \text{ F/m}$$

Determine its π equivalent model.

5.13. A three-phase transmission line has π equivalent parameters as shown below. Its receiving end has a load of $100 + j700$ Ω per phase, and its sending end has a voltage of 230 kV. Calculate the following quantities:
(a) sending end current per phase
(b) sending end power per phase
(c) receiving end voltage

5.9 PROBLEMS

(d) receiving end current per phase
(e) receiving end power per phase
Line parameters are
  series resistance = 20 Ω
  series reactance = $j120$ Ω
  shunt capacitive admittance = $j0.0012$ ℧

5.14. Repeat Problem 5.13 for a load of $100 - j500$ Ω.

5.15. Repeat Problem 5.13 for no load at the receiving end of the line.

5.16. Calculate the line-to-neutral capacitance of the three-phase line in Problem 5.5 while considering ground effects. Assume conductor 3 is 5 m aboveground.

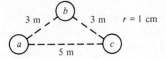

Figure 5.39 Transmission line.

5.17. An untransposed three-phase transmission line has a cross section as shown in Figure 5.39. It is being converted to a single-phase line such that conductor 1 is the parallel combination of phase a and b while conductor 2 is phase c. Calculate L_1 and L_2 of the single-phase line.

Chapter 6

Network Analysis

6.1 INTRODUCTION

Due to environmental, economic, technological, and reliability concerns, generating electric energy at the same location it is used is not feasible. An extensive system has been constructed to transmit energy which has been generated at one location and is used at several locations many miles away.

Large amounts of power are generated at central generating stations and sent to a network of high-voltage transmission lines through step-up transformers. These transmission lines supply the power to lower voltage subtransmission networks, which supply power to still lower voltage distribution networks. Each of the three networks is connected to the others through step-down transformers. The subtransmission and distribution network lines supply power to customers directly. Thus, the total network is a complex grid of interconnected lines. This network has the function of transmitting power from the points of generation to the points of consumption. The importance of this function requires engineering analysis of key network parameters under various conditions. These parameters include voltages, power flows, and power losses. This chapter is devoted to illustrating methods for performing such an analysis.

6.2 ONE-LINE DIAGRAMS AND IMPEDANCE DIAGRAMS

A balanced three-phase system is almost always analyzed as a single-phase circuit. This kind of analysis means that the system is represented by a diagram of one phase with an assumed neutral return. A single-phase representation

6.2 ONE-LINE DIAGRAMS AND IMPEDANCE DIAGRAMS

for balanced three-phase operation may be used because the electrical parameters calculated for one phase will be exactly the same in the other two phases but with a $\pm 120°$ phase shift. A diagram of a network is simplified by representing its elements by symbols rather than by equivalent circuits. Such a diagram is called a one-line diagram. Figure 6.1 illustrates such a diagram.

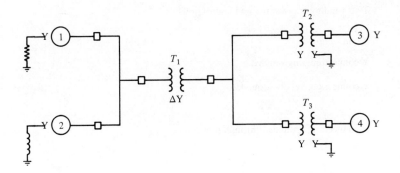

Generators 1 and 2: 30 MVA, 10 kV; $X = 0.8$ Ω

Synchronous motors 3 and 4: 20 MVA, 10 kV; $X = 1.1$ Ω

T_1, T_2, T_3: three single-phase transformers in each bank; each transformer 10 MVA, 10/150 kV; $Z = 1.2 + j12$ Ω referred to high-voltage side

Impedance of transmission lines $= 2 + j20$ Ω

Figure 6.1 One-line diagram.

The symbols of Figure 6.1 are explained in Table 6.1. Additional symbols may be added if they are needed. Transmission lines are represented as straight lines, such as those connecting the transformers in Figure 6.1. Element data are usually given not on the drawing but as separate information.

In order to perform calculations on the system in Figure 6.1, the symbols must be replaced by single-phase equivalent circuits of the elements. On the assumption that transformer magnetization currents are negligible, transformers can be represented as series impedances. Assuming that all transmission lines are short, they may be represented as a resistance and inductive reactance in series. Synchronous machine symbols can be replaced with impedances and voltage sources. Figure 6.2 illustrates the system in Figure 6.1 with all values referred to the common side of the transformers. Figure 6.2 is an impedance diagram of the one-line diagram of Figure 6.1. Notice that impedances of the grounding circuits of the generators are not included in the impedance diagram. Under balanced three-phase conditions no current would flow in the neutral, so the neutral impedance would have no effect on the system.

TABLE 6.1

Symbol	Explanation
Y ◯	Rotating machine and connection
▢	Power circuit breaker
⅊⅌ ΔY	Two-winding transformer and connection
Δ	Δ connection
Y	Y connection, ungrounded
	Y connection, solidly grounded
	Y connection, grounded through inductance
	Y connection, grounded through resistance

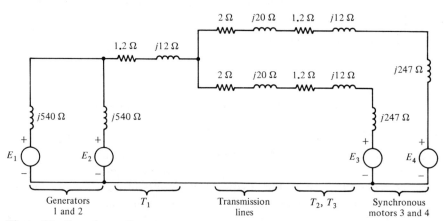

Figure 6.2 Impedance diagram.

6.3 PER UNIT QUANTITIES

Networks can be analyzed using volts, amperes, ohms, and voltamperes. However, referring these quantities to a per unit (pu) or percent value of base voltage, current, impedance, or voltamperes can make calculations much easier.

6.3 PER UNIT QUANTITIES

The per unit method expresses a quantity as a decimal fraction of a chosen base quantity. For example, for a chosen base of 10,000 V, a voltage of 9000 V would have a per unit value of 0.9 pu, while 11,000 V would be 1.1 pu. Per unit quantities are determined by dividing the actual value by the base value:

$$\text{Per unit value} = \frac{\text{actual value}}{\text{base value}} \tag{6.1}$$

Percent quantities are obtained by multiplying per unit values by 100. The per unit method of calculation has an advantage over the percent method in that the product of two per unit values is a per unit value. The product of two percent values must be divided by 100 to obtain the correct percent result. Therefore, per unit calculations are favored over percent. However, impedances are often given in percent and must be converted to per unit for calculations. Care should be taken in noting if given quantities are in per unit or percent in order to make proper conversions before beginning calculations.

All relationships between voltage, current, impedance, and voltamperes must be maintained when using per unit analysis. If base values of any two of these network quantities are specified, then base values for the other two are fixed. Usually, base volts and base voltamperes are the quantities chosen to be specified. For a single-phase system the base current is then base voltamperes divided by base volts. The base impedance is also fixed as base voltage divided by base current.

$$\text{Base } I = \frac{\text{base } VA}{\text{base } V} \tag{6.2}$$

$$\text{Base } Z = \frac{\text{base } V}{\text{base } I}$$

$$= \frac{(\text{base } V)^2}{\text{base } VA} \tag{6.3}$$

Most power networks are composed of balanced three-phase circuits. These balanced three-phase circuits are commonly solved as single-phase circuits with a neutral return. However, system data are usually given and results desired in total three-phase voltamperes and line-to-line voltages. The direct application of Equations 6.2 and 6.3 to the solution of the three-phase circuits by single-phase representation requires that line-to-neutral voltage and single-phase voltampere bases be chosen. Therefore, the analysis requires that the following steps be taken:

1. Convert known system line-to-line voltages and three-phase powers to line-to-neutral and single-phase quantities.
2. Determine per unit values of the line-to-neutral voltages and single-phase powers.
3. Do a per unit analysis.

4. Convert the per unit results to line-to-neutral and single-phase quantities using the per unit bases.
5. Convert the results of step 4 to line-to-line voltages and three-phase powers.

EXAMPLE 6.1

For the three-phase system shown, the line-to-line voltage at bus 2 is 19 kV. The three-phase load at bus 2 is $300 + j120$ kVA. The transmission line connecting buses 1 and 2 has an impedance of $17.2 + j82.2$ Ω/phase. Using per unit analysis determine the line-to-line voltage at bus 1. Use a line-to-neutral voltage base of 10.0 kV and a single-phase voltampere base of 90.0 kVA.

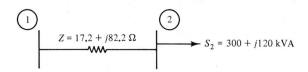

The given and calculated base quantities are as follows:

$$\text{Base } V_{l\text{-}n} = 10.0 \text{ kV}$$
$$\text{Base } VA_{1\phi} = 90.0 \text{ kVA}$$
$$\text{Base line current} = \frac{\text{base } VA_{1\phi}}{\text{base } V_{l\text{-}n}}$$
$$= \frac{90.0 \times 10^3}{10.0 \times 10^3}$$
$$= 9.0 \text{ A}$$
$$\text{Base impedance} = \frac{(\text{base } V_{l\text{-}n})^2}{\text{base } VA_{1\phi}}$$
$$= \frac{(10.0 \times 10^3)^2}{90.0 \times 10^3}$$
$$= 1.11 \times 10^3 \ \Omega$$

The subscripts $l\text{-}n$ and 1ϕ indicate that the quantities are line-to-neutral and single phase, respectively. The voltage at bus 2 must be converted to a line-to-neutral quantity and the load to a single-phase quantity:

$$V_{2,l\text{-}n} = \frac{V_{2,l\text{-}l}}{\sqrt{3}}$$
$$= \frac{19 \times 10^3}{\sqrt{3}}$$
$$= 11.0 \text{ kV}$$

6.3 PER UNIT QUANTITIES

$$S_{2,1\phi} = \frac{S_{2,3\phi}}{3}$$

$$= \frac{(300 + j120) \times 10^3}{3}$$

$$= 100 + j40 \text{ kVA}$$

The subscripts *l-l* and 3ϕ indicate that the quantities are line-to-line and three phase, respectively. If the line-to-neutral voltage at bus 2 is chosen as reference, the per unit voltage and load at bus 2 can be determined:

$$V_{2,pu} = \frac{\text{actual } V_{2,l-n}}{\text{base } V_{l-n}}$$

$$= \frac{11.0 \times 10^3 \underline{/0°}}{10.0 \times 10^3}$$

$$= 1.10\underline{/0°} \text{ pu}$$

$$S_{2,pu} = \frac{\text{actual } S_{2,1\phi}}{\text{base } VA_{1\phi}}$$

$$= \frac{(100 + j40) \times 10^3}{90.0 \times 10^3}$$

$$= 1.20\underline{/21.8°} \text{ pu}$$

The per unit line current can now be calculated from $S = VI^*$:

$$I_{pu} = \left(\frac{S_{2,pu}}{V_{2,pu}}\right)^*$$

$$= \left(\frac{1.20\underline{/21.8°}}{1.10\underline{/0°}}\right)^*$$

$$= 1.09\underline{/-21.8°} \text{ pu}$$

The per unit impedance is found from the base impedance:

$$Z_{pu} = \frac{\text{actual } Z}{\text{base } Z}$$

$$= \frac{17.2 + j82.2}{1.11 \times 10^3}$$

$$= 0.076\underline{/78.2°} \text{ pu}$$

Using Kirchhoff's voltage law, the per unit voltage at bus 1 can be determined:

$$V_{1pu} = V_{2pu} + I_{pu}Z_{pu}$$
$$= 1.10\underline{/0°} + (1.09\underline{/-21.8°})(0.076\underline{/78.2°})$$
$$= 1.15\underline{/3.4°} \text{ pu}$$

The line-to-neutral voltage at bus 1 in volts can be calculated from the base voltage:

$$\begin{aligned} V_{1,l\text{-}n} &= V_{1\mathrm{pu}}(\text{base } V_{l\text{-}n}) \\ &= 1.15\underline{/3.4°}(10.0 \times 10^3) \\ &= 11.5\underline{/3.4°} \text{ kV} \end{aligned}$$

The line-to-line voltage magnitude at bus 1 can now be determined:

$$\begin{aligned} V_{1,l\text{-}l} &= \sqrt{3}V_{1,l\text{-}n} \\ &= \sqrt{3}(11.5 \times 10^3) \\ &= 19.9 \text{ kV} \end{aligned}$$ ■■

At the beginning of Example 6.1, all system voltages and voltamperes were converted from line-to-line and three-phase quantities to line-to-neutral and single-phase values. These values were then converted to per unit. At the end of the problem with the answer in per unit, this process was reversed to obtain the answer in line-to-line volts. Since converting from three-phase or line-to-line to per unit consists of two divisions by constants, simplifying the procedure by combining these operations is desirable. Similarly, the two multiplications to convert back from per unit should also be combined. The per unit voltage and voltamperes could be calculated as follows:

$$V_{\mathrm{pu}} = \frac{\text{actual } V_{l\text{-}l}}{\sqrt{3}(\text{base } V_{l\text{-}n})} \qquad (6.4)$$

$$VA_{\mathrm{pu}} = \frac{\text{actual } VA_{3\phi}}{3(\text{base } VA_{1\phi})} \qquad (6.5)$$

The denominator of Equation 6.4 is the line-to-line equivalent of the base line-to-neutral voltage. Similarly, the denominator of Equation 6.5 is the three-phase equivalent of the base single-phase voltamperes. Therefore, Equations 6.4 and 6.5 can be written in line-to-line and three-phase form:

$$V_{\mathrm{pu}} = \frac{\text{actual } V_{l\text{-}l}}{\text{base } V_{l\text{-}l}} \qquad (6.6)$$

$$VA_{\mathrm{pu}} = \frac{\text{actual } VA_{3\phi}}{\text{base } VA_{3\phi}} \qquad (6.7)$$

Note that these per unit quantities will still be used in a single-phase representation of three-phase circuits. Even though three-phase voltamperes and line-to-line voltages are used to calculate per unit values, single-phase calculations must be used in the per unit network analysis. For example, $S_{\mathrm{pu}} = V_{\mathrm{pu}}I_{\mathrm{pu}}^*$ is the equation for determining per unit complex power being supplied by per unit voltage V and per unit line current I.

In order for the per unit line currents and per unit impedances to have the same values with the line-to-line and three-phase bases as they had with the line-to-neutral and single-phase bases, the base current and base impedance

6.3 PER UNIT QUANTITIES

calculated from each set of voltage and voltamperes bases must be the same. Starting with Equations 6.2 and 6.3, these quantities are derived for line-to-line voltage base and three-phase voltampere base:

$$\text{Base } I = \frac{\text{base } VA_{1\phi}}{\text{base } V_{l\text{-}n}}$$

$$= \frac{\text{base } VA_{3\phi}/3}{\text{base } V_{l\text{-}l}/\sqrt{3}}$$

$$= \frac{\text{base } VA_{3\phi}}{\sqrt{3}(\text{base } V_{l\text{-}l})} \tag{6.8}$$

$$\text{Base } Z = \frac{(\text{base } V_{l\text{-}n})^2}{\text{base } VA_{1\phi}}$$

$$= \frac{(\text{base } V_{l\text{-}l}/\sqrt{3})^2}{\text{base } VA_{3\phi}/3}$$

$$= \frac{(\text{base } V_{l\text{-}l})^2}{\text{base } VA_{3\phi}} \tag{6.9}$$

These bases can be used to convert line-to-line voltages, three-phase voltamperes, line currents, and per phase impedances to per unit quantities without first converting to line-to-neutral and single-phase values. The line-to-line and three-phase set of bases is the most frequently used for three-phase circuits. When a voltage and voltampere are given as bases for three-phase circuits, they are assumed to be line-to-line and three-phase quantities unless otherwise stated.

EXAMPLE 6.2

Repeat Example 6.1 using a voltage base of 17.32 kV and a voltampere base of 270 kVA. The given and calculated bases are

$$\text{Base } V = 17.32 \text{ kV}$$
$$\text{Base } VA = 270 \text{ kVA}$$

$$\text{Base line current} = \frac{\text{base } VA}{\sqrt{3}(\text{base } V)}$$

$$= \frac{270 \times 10^3}{\sqrt{3}(17.32 \times 10^3)}$$

$$= 9.0 \text{ A}$$

$$\text{Base impedance} = \frac{(\text{base } V)^2}{\text{base } VA}$$

$$= \frac{(17.32 \times 10^3)^2}{270 \times 10^3}$$

$$= 1.11 \times 10^3 \text{ }\Omega$$

The per unit voltage and load at bus 2 are calculated directly from given data:

$$V_{2\text{pu}} = \frac{\text{actual } V_{2,l\text{-}l}}{\text{base } V}$$

$$= \frac{19.0 \times 10^3}{17.32 \times 10^3}$$

$$= 1.10 \text{ pu}$$

$$S_{2\text{pu}} = \frac{\text{actual } S_{2,3\phi}}{\text{base } V}$$

$$= \frac{(300 + j120) \times 10^3}{270 \times 10^3}$$

$$= 1.20\underline{/21.8°} \text{ pu}$$

Choosing $V_{2\text{pu}}$ as reference, the per unit line current is calculated:

$$I_{\text{pu}} = \left(\frac{S_{2\text{pu}}}{V_{2\text{pu}}}\right)^*$$

$$= \left(\frac{1.20\underline{/21.8°}}{1.10\underline{/0°}}\right)^*$$

$$= 1.09\underline{/-21.8°} \text{ pu}$$

The per unit impedance is determined and $V_{1\text{pu}}$ calculated:

$$Z_{\text{pu}} = \frac{\text{actual } Z}{\text{base } Z}$$

$$= \frac{17.2 + j82.2}{1.11 \times 10^3}$$

$$= 0.076\underline{/78.2°} \text{ pu}$$

$$V_{1\text{pu}} = V_{2\text{pu}} + I_{\text{pu}} Z_{\text{pu}}$$
$$= 1.10\underline{/0°} + (1.09\underline{/-21.8°})(0.076\underline{/78.2°})$$
$$= 1.15\underline{/3.4°} \text{ pu}$$

The line-to-line voltage at bus 1 is calculated directly:

$$V_1 = V_{1\text{pu}}(\text{base } V)$$
$$= (1.15)(17.32 \times 10^3)$$
$$= 19.9 \text{ kV} \qquad \blacksquare\blacksquare$$

6.3.1 Changing per Unit Bases

The impedance of a device is usually given in per unit in terms of that device's voltage and voltampere ratings. If the device is to be incorporated in a system analysis, all of its given per unit values must be changed to values in the system per unit bases. A per unit impedance given on impedance base 1 can be con-

6.3 PER UNIT QUANTITIES

verted to a per unit impedance on impedance base 2 in the following manner. First, change the per unit impedance on base 1 to an actual ohmic impedance:

$$\text{Actual } Z = Z_{pu1} \frac{(\text{base } V_1)^2}{\text{base } VA_1} \tag{6.10}$$

Next, convert the actual impedance to a per unit value on base 2:

$$Z_{pu2} = \text{actual } Z \bigg/ \frac{(\text{base } V_2)^2}{\text{base } VA_2} \tag{6.11}$$

Equations 6.10 and 6.11 can be combined to give a one-step conversion equation:

$$Z_{pu2} = Z_{pu1} \left(\frac{\text{base } V_1}{\text{base } V_2} \right)^2 \left(\frac{\text{base } VA_2}{\text{base } VA_1} \right) \tag{6.12}$$

EXAMPLE 6.3

A three-phase generator rated at 720 MVA and 20 kV has a per phase reactance of 0.35 pu on its own base. The generator is to be placed in a system where the bases are 100 MVA and 13.8 kV. Find the reactance of the generator in per unit on the new base:

$$X_{pu2} = X_{pu1} \left(\frac{\text{base } V_1}{\text{base } V_2} \right)^2 \left(\frac{\text{base } VA_2}{\text{base } VA_1} \right)$$

$$= 0.35 \left(\frac{20 \times 10^3}{13.8 \times 10^3} \right)^2 \left(\frac{100 \times 10^6}{720 \times 10^6} \right)$$

$$= 0.102 \text{ pu} \qquad \blacksquare\blacksquare$$

6.3.2 Determining per Unit Bases

As was shown in Chapter 4, the ohmic value of impedance of a single-phase transformer is dependent on whether it is measured on the high-voltage side or the low-voltage side of the device. The per unit impedance of the transformer based on its own ratings may be determined from either the high- or low-voltage side. If the ohmic impedance is measured from the low-voltage side, then the low-side voltage rating of the transformer must be used as the base voltage. Similarly, if the impedance is measured from the high-voltage side, then the high side voltage rating is the base. In either case the voltampere rating of the transformer is the voltampere base. The per unit impedance has the same value regardless of which side of the transformer it is calculated from.

EXAMPLE 6.4

A single-phase transformer is rated at 220/4000 V and 5.0 kVA. The reactance of the transformer is 0.1 Ω measured from the low-voltage side. Determine the per unit reactance of the transformer on its own base from both the high- and low-voltage sides.

The impedance base on the low-voltage side of the transformer would be

$$\text{Base } Z_L = \frac{(\text{base } V_L)^2}{\text{base } VA}$$

$$= \frac{(220)^2}{5.0 \times 10^3} \, \Omega$$

The subscript L indicates low-voltage side. The per unit reactance calculated from the low-voltage side values is

$$X_{Lpu} = \frac{\text{actual } X_L}{\text{base } Z_L}$$

$$= 0.1 \bigg/ \frac{(220)^2}{5.0 \times 10^3}$$

$$= 0.1 \left[\frac{5.0 \times 10^3}{(220)^2} \right]$$

$$= 0.0103 \text{ pu}$$

The ohmic value of reactance measured on the high-voltage side is related to the ohmic value measured on the low-voltage side by the ratio of transformation squared:

$$\text{Actual } X_H = \text{actual } X_L(a^2)$$

$$= 0.1 \left(\frac{4000}{220} \right)^2 \, \Omega$$

The subscript H indicates high-voltage side values. The base impedance on the high-voltage side would be

$$\text{Base } Z_H = \frac{(\text{base } V_H)^2}{\text{base } VA}$$

$$= \frac{(4000)^2}{5.0 \times 10^3} \, \Omega$$

The per unit impedance calculated from the high-voltage side values is

$$X_{Hpu} = \frac{\text{Actual } X_H}{\text{Base } Z_H}$$

$$= 0.1 \left(\frac{4000}{220} \right)^2 \bigg/ \frac{(4000)^2}{5.0 \times 10^3}$$

$$= 0.1 \left(\frac{5.0 \times 10^3}{(220)^2} \right)$$

$$= 0.0103 \text{ pu} \qquad \blacksquare\blacksquare$$

When using the transformer's ratings as base, the per unit impedance is the same value regardless of which side of the transformer it is calculated from.

6.3 PER UNIT QUANTITIES

This conclusion is true for bases other than the transformer's own base. The only requirement is that the ratio of the base voltages used on the high- and low-voltage sides be equal to the ratio of transformation of the transformer. In addition, the voltampere base must be the same on both sides.

EXAMPLE 6.5

Repeat Example 6.4 using a voltampere base of 10 kVA and a high-voltage side base of 800 V.

The ratio of voltage bases must be equal to the turns ratio. The low-voltage side base can be calculated:

$$800\left(\frac{220}{4000}\right) = 44 \text{ V}$$

On the low-voltage side the base impedance and per unit reactance are

$$\text{Base } Z_L = \frac{(\text{base } V_L)^2}{\text{base } VA}$$

$$= \frac{(44)^2}{10,000}$$

$$= 0.1936 \, \Omega$$

$$X_{Lpu} = \frac{\text{actual } X_L}{\text{base } Z_L}$$

$$= \frac{0.1}{0.1936}$$

$$= 0.517 \text{ pu}$$

The ohmic value of reactance measured on the high-voltage side of the transformer was found in Example 6.4. The high-voltage side base impedance and per unit reactance are

$$\text{Base } Z_H = \frac{(\text{base } V_H)^2}{\text{base } VA}$$

$$= \frac{(800)^2}{10,000}$$

$$= 64.0 \, \Omega$$

$$X_{Hpu} = \frac{\text{actual } X_H}{\text{base } Z_H}$$

$$= 0.1 \frac{(4000/220)^2}{64.0} = 0.517 \text{ pu} \quad \blacksquare\blacksquare$$

By choosing voltage bases on each side of a transformer so that their ratio is the turns ratio of the transformer and by choosing a common voltampere base, the per unit value of an impedance will be the same on either side of a transformer. Therefore, if voltage and voltampere bases are chosen for one part of a system, the bases are fixed for the rest of the system. When this practice is employed, the per unit impedance values calculated on different bases in different parts of the system can be assembled and used on a single-impedance diagram for the whole system.

EXAMPLE 6.6

A line-to-neutral voltage base of 7 kV and a single-phase voltampere base of 100 kVA were chosen for section A of the single-phase system shown below. The transformer is rated at 8750/500 V and 150 kVA. The transformer has a leakage reactance of 10 percent. Draw a per unit impedance diagram of the system including the voltage source and determine the voltage across the load in section B. The bases must first be determined.

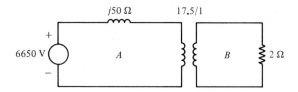

For section A,

$$\text{Base } V = 7 \text{ kV}$$
$$\text{Base } VA = 100 \text{ kVA}$$
$$\text{Base } Z = \frac{(7 \times 10^3)^2}{100 \times 10^3} = 490 \ \Omega$$

For section B,

$$\text{Base } V = 7 \times 10^3 \left(\frac{500}{8750}\right) = 400 \text{ V}$$
$$\text{Base } VA = 100 \text{ kVA}$$
$$\text{Base } Z = \frac{(400)^2}{100 \times 10^3} = 1.6 \ \Omega$$

The per unit reactance of the transformer must be calculated on the system base:

$$\text{Per unit transformer reactance} = 0.1 \left(\frac{8.75 \times 10^3}{7 \times 10^3}\right)^2 \left(\frac{100 \times 10^3}{150 \times 10^3}\right)$$
$$= 0.104 \text{ pu}$$

Next, the ohmic impedances in sections A and B must be converted to per unit:

$$\text{Per unit impedance of section } A = \frac{j50}{490}$$
$$= j0.102 \text{ pu}$$

6.3 PER UNIT QUANTITIES

$$\text{Per unit impedance of section } B = \frac{2}{1.6}$$

$$= 1.25 \text{ pu}$$

The voltage of the source in section A must also be in per unit:

$$\text{Per unit voltage of source} = \frac{6650}{7000}$$

$$= 0.95 \text{ pu}$$

The per unit impedance diagram can now be drawn:

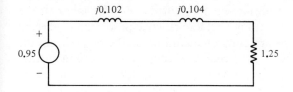

By applying voltage division and choosing the voltage source as reference, the per unit voltage across the load of section B can be determined:

$$\text{Per unit voltage across the load} = 0.95\underline{/0°}\left(\frac{1.25}{1.25 + j0.102 + j0.104}\right)$$

$$= 0.937\underline{/-9.4°} \text{ pu}$$

The actual voltage is found from the base voltage of section B:

$$\text{Voltage across load} = 0.937(400)$$

$$= 375 \text{ V} \qquad \blacksquare\blacksquare$$

In applying the above principle to three-phase circuits, care must be taken in determining the ratio of the voltage bases on each side of the transformer bank. The voltage bases must be in the same ratio as the rated line-to-line voltages of each side of the bank. If the transformer bank consists of a three-phase transformer, then the ratings of the device are given in line-to-line voltages and three-phase voltamperes. The ratio of the voltage bases is given by the ratio of the rated line-to-line voltages of a three-phase transformer.

If the three-phase transformer bank consists of three identical single-phase transformers connected in Δ-Δ or Y-Y, the ratio of voltage bases is again easily found. For the Δ-Δ connection the voltage rating on each transformer would be the line-to-line rating of the bank. Therefore, the ratio of transformation for each transformer would be the ratio required for the voltage bases. If the bank is connected Y-Y, the voltage rating of each transformer is the line-to-neutral voltage rating of the bank. On both sides of the bank the line-to-neutral voltage is related to the line-to-line voltage by $\sqrt{3}$. The ratio of line-to-line voltages is, therefore, the same as the ratio of line-to-neutral voltages. The ratio of transformation for each transformer would also be the ratio required for the voltage bases.

If the three single-phase transformers are connected in Δ-Y, choosing the correct voltage bases becomes slightly more complicated. In this case the voltage ratings of the transformers would represent a line-to-line voltage on one side and a line-to-neutral voltage on the other. To obtain a ratio of line-to-line voltages for the bank, the voltage rating corresponding to the Y side of the bank must be multiplied by $\sqrt{3}$. The ratio of the Δ side rating of each transformer to the Y side rating multiplied by $\sqrt{3}$ will be the ratio required for the voltage bases.

It is important to remember at this point that the Δ-Y and Y-Δ connections also have a 30° phase shift associated with them. However, since all voltages and currents across the transformer are shifted the same amount, the magnitudes of voltages, currents, real power, and reactive power are unaffected by this shift. The phase relationships between quantities on the same side of the transformer are also unaffected. It is, therefore, a common practice in steady-state analysis to work with just the equivalent Y circuit for these connections and avoid the 30° phase shift. If, however, the actual phase relationship between voltages or currents on different sides of a Δ-Y- or Y-Δ-connected transformer bank is needed, then the 30° phase shift must be taken into account.

6.3.3 Per Unit Single-Phase Impedance Representation of Δ-Connected Devices

The per unit value of impedance of a three-phase transformer bank is unchanged whether the single-phase transformers are connected in Y or Δ. This situation can be illustrated by analyzing a single-phase device rated at 17 kV and 10 kVA and having an impedance of 10 percent. The device is connected in Y with two other identical devices. Using the device's ratings, the system bases are $17 \times \sqrt{3}$ kV and 30 kVA. The single-phase impedance representation of the three devices would have a value of 0.1 pu. What would that single-phase representation be if the devices were connected in Δ? For the Δ connection the system bases would be 17 kV and 30 kVA. Therefore, the 10 percent impedance must be converted to a new voltage base:

$$0.1 \left(\frac{17(\sqrt{3}) \times 10^3}{17 \times 10^3} \right)^2 = 0.3 \text{ pu}$$

However, to represent the three-phase Y connection as a single phase with neutral return, a Y-Δ transformation must be performed. For a balanced Δ connection this is accomplished by dividing the Δ phase impedances by three:

$$0.3/3 = 0.1 \text{ pu}$$

Therefore, whether a device is Δ or Y connected has no effect on its per unit impedance value. This reasoning can be extended to any combination of three-phase transformer connections.

6.3 PER UNIT QUANTITIES

EXAMPLE 6.7

Draw an impedance diagram of the system shown below including per unit loads using the generator bases as system bases in that part of the network:

Generator:	30 MVA 15 kV	$X = 17$ percent
Transformer T_1:	30 MVA 15Δ/138Y kV	$X = 10$ percent
Transformer T_2:	three single-phase transformers each rated at 8 MVA 8/80 kV $X = 10$ percent	
Transformer T_3:	three single-phase transformers each rated at 10 MVA 9/81 kV $X = 10$ percent	

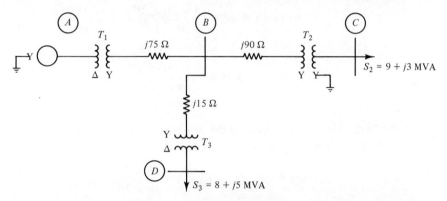

First, the bases must be determined for all sections of the network. The bases in section A are given as the generator's bases:

$$\text{Section } A \text{ voltage base} = 15 \text{ kV}$$
$$\text{Section } A \text{ voltampere base} = 30 \text{ MVA}$$

Section A voltage data fixes the voltage bases in the remaining sections of the system. The voltampere base is 30 MVA in all sections. Section B is connected to section A by a three-phase transformer, so the voltage base in section B is

$$15 \times 10^3 \left(\frac{138 \times 10^3}{15 \times 10^3} \right) = 138 \text{ kV}$$

Section C is connected to section B by three single-phase Y-Y transformers, so section C base voltage is

$$138 \times 10^3 \left[\frac{(\sqrt{3})8.0 \times 10^3}{(\sqrt{3})80 \times 10^3} \right] = 13.8 \text{ kV}$$

Section D is connected to section B by three single-phase Δ-Y-connected transformers, so section D base voltage is

$$138 \times 10^3 \left[\frac{9.0 \times 10^3}{(\sqrt{3})81 \times 10^3} \right] = 8.85 \text{ kV}$$

The system base at the generator is the generator's base so the per unit impedance of the generator does not have to be modified. The same is true for transformer T_1. However, the impedances of transformers T_2 and T_3 must be converted to the system bases by using Equation 6.12:

$$T_2: \quad 0.1\left(\frac{\sqrt{3}\,80 \times 10^3}{138 \times 10^3}\right)^2 \left(\frac{30 \times 10^6}{(3)8 \times 10^6}\right) = 0.126 \text{ pu}$$

$$T_3: \quad 0.1\left(\frac{\sqrt{3}\,81 \times 10^3}{138 \times 10^3}\right)^2 = 0.103 \text{ pu}$$

The loads must also be represented as per unit values:

$$\text{Section } C \text{ load} = \frac{(9+j3) \times 10^6}{30 \times 10^6}$$

$$= 0.3 + j0.1 \text{ pu}$$

$$\text{Section } D \text{ load} = \frac{(8+j5) \times 10^6}{30 \times 10^6}$$

$$= 0.27 + j0.17 \text{ pu}$$

All the ohmic impedances given in this system are in section B. Therefore, the impedance base for section B must be calculated:

$$\text{Section } B \text{ impedance base} = \frac{(138 \times 10^3)^2}{30 \times 10^6}$$

$$= 635 \ \Omega$$

Now the ohmic impedances can be converted to per unit and used to draw the impedance diagram:

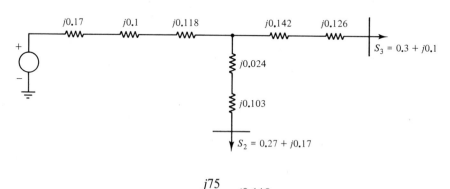

$$\frac{j75}{635} = j0.118 \text{ pu}$$

$$\frac{j90}{635} = j0.142 \text{ pu}$$

$$\frac{j15}{635} = j0.024 \text{ pu}$$

■■

6.3.4 The Advantages of per Unit Versus Ohmic Representation

Per unit impedances of machines of the same type usually fall within a narrow range when given on the machine's base. The ohmic values, however, vary widely with the ratings of the machines. Therefore, if the impedance is unavailable for a particular device, an appropriate per unit value can usually be estimated.

When ohmic values are used in an equivalent circuit, all values must be referred to one part of the circuit. This process is performed by referring each impedance through the transformers connecting that impedance to the reference part of the circuit. The per unit method results in the same values for impedances on either side of a transformer. Therefore, when proper bases are chosen, no impedances must be referred through transformers.

The per unit impedance of a three-phase transformer bank is not affected by the way in which the transformers are connected. However, the relationship between the voltage bases on each side of the bank is determined by the connection.

The per unit method greatly simplifies the work required to analyze a system. A greater appreciation of the method will come as it is used.

6.4 POWER FLOW ANALYSIS

Several computational methods have been developed to help engineers analyze the way in which a system responds to certain loads and generation schedules. These methods are used to answer such questions as these:

1. How does the real and reactive power flow through the system under certain conditions?
2. Are any transmission lines, transformers, or other equipment overloaded?
3. What is the voltage at all buses?

The type of analysis used to answer these questions is called power flow or load flow and is not a straightforward linear circuit problem. Complications arise because loads on a power system usually behave like constant power sinks and cannot be represented as constant impedances. Also, generator EMFs, E_f, are usually not known. Therefore, a method of analyzing power systems will be developed in this section that accommodates these kinds of characteristics into the power flow solution.

6.4.1 Bus Admittance Matrix

With a system represented in per unit quantities, a solution can be found for the unknown parameters of that system using well-known circuit procedures.

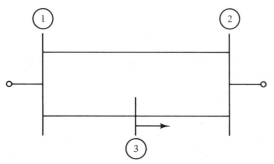

Figure 6.3 Simple power system one-line diagram.

One of the most useful methods of solution for power systems is the application of Kirchhoff's current law in writing nodal equations. This method of analysis will be developed with reference to the system in Figure 6.3. If the system impedances are known, an impedance diagram of Figure 6.3 could be drawn. However, using admittances is more convenient for writing nodal equations. Therefore, the diagram in Figure 6.4 has admittances indicated. These admittances are the reciprocals of their corresponding impedances.

Nodal equations can now be written for the system. The current entering a node from a generator or load is equal to the current leaving that node through all other system elements. For node 1 Kirchhoff's current law gives

$$I_1 = (V_1 - V_2)Y_c + (V_1 - V_3)Y_d \qquad (6.13)$$

Current I_1 represents that current that is entering node 1 from the generator at node 1. For node 2

$$I_2 = (V_2 - V_1)Y_c + (V_2 - V_3)Y_e \qquad (6.14)$$

The current entering node 2 from its generator is represented by I_2. For node 3

$$I_3 = (V_3 - V_1)Y_d + (V_3 - V_2)Y_e \qquad (6.15)$$

The left side of Equation 6.15 represents the current entering node 3 from the load at node 3. Since loads normally take current from the system instead of

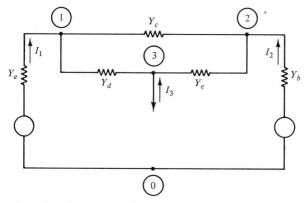

Figure 6.4 Admittance diagram.

6.4 POWER FLOW ANALYSIS

supplying it, I_3 would have a negative value. Voltage V_1, V_2, and V_3 are defined as the voltages from bus 1, 2, and 3 to the reference node, respectively.

If I_1, I_2, and I_3 are known or can be calculated, Equations 6.13 to 6.15 are three independent equations with three unknowns. Therefore, a solution for values of V_1, V_2, and V_3 can be found, and using these voltages all branch currents can be determined. This conclusion implies that the number of required nodal equations needed to solve any circuit is one less than the number of nodes in that circuit.

If Equations 6.13 to 6.15 are rearranged so as to collect the coefficients of each variable V_1, V_2, and V_3, they yield

$$\begin{aligned} I_1 &= (Y_c + Y_d)V_1 + (-Y_c)V_2 + (-Y_d)V_3 \\ I_2 &= (-Y_c)V_1 + (Y_c + Y_e)V_2 + (-Y_e)V_3 \\ I_3 &= (-Y_d)V_1 + (-Y_e)V_2 + (Y_d + Y_e)V_3 \end{aligned} \quad (6.16)$$

Equations 6.16 can be written in the form

$$\begin{aligned} I_1 &= Y_{11}V_1 + Y_{12}V_2 + Y_{13}V_3 \\ I_2 &= Y_{21}V_2 + Y_{22}V_2 + Y_{23}V_3 \\ I_3 &= Y_{31}V_3 + Y_{32}V_2 + Y_{33}V_3 \end{aligned} \quad (6.17)$$

Admittances Y_{11}, Y_{22}, and Y_{33} of Equations 6.17 are self-admittances of each node and equal the sum of all admittances except generators and loads connected to that node. Admittances Y_{12}, Y_{13}, Y_{21}, Y_{23}, Y_{31}, and Y_{32} are mutual admittances. The mutual admittances are equal to the negative of the sum of the admittances directly connecting the nodes identified by their subscripts.

Equations 6.17 can be written in matrix form. A brief review of matrix operations is given in Appendix C.

$$\begin{bmatrix} I_1 \\ I_2 \\ I_3 \end{bmatrix} = \begin{bmatrix} Y_{11} & Y_{12} & Y_{13} \\ Y_{21} & Y_{22} & Y_{23} \\ Y_{31} & Y_{32} & Y_{33} \end{bmatrix} \begin{bmatrix} V_1 \\ V_2 \\ V_3 \end{bmatrix} \quad (6.18)$$

The square matrix of self and mutual admittances in Equation 6.18 is called the bus admittance matrix.

EXAMPLE 6.8

For the system shown below, the known system data are as follows in per unit.

$$Z_a = j0.5 \quad Z_c = j1.2 \quad Z_e = j0.25$$
$$Z_b = j0.25 \quad Z_d = j1.5$$

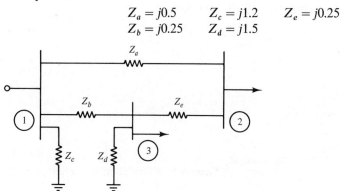

Find the bus admittance matrix for this system. The admittances are

$$Y_a = -j2.0 \qquad Y_c = -j0.83 \qquad Y_e = -j4.0$$
$$Y_b = -j4.0 \qquad Y_d = -j0.67$$

The self-admittances are found to be

$$Y_{11} = Y_a + Y_b + Y_c$$
$$= -j6.83$$
$$Y_{22} = Y_a + Y_e$$
$$= -j6.0$$
$$Y_{33} = Y_b + Y_d + Y_e$$
$$= -j8.67$$

The mutual admittances are

$$Y_{12} = Y_{21}$$
$$= -Y_a$$
$$= j2.0$$
$$Y_{13} = Y_{31}$$
$$= -Y_b$$
$$= j4.0$$
$$Y_{23} = Y_{32}$$
$$= -Y_e$$
$$= j4.0$$

The admittances can now be assembled into the bus admittance matrix

$$\begin{bmatrix} -j6.83 & j2.0 & j4.0 \\ j2.0 & -j6.0 & j4.0 \\ j4.0 & j4.0 & -j8.67 \end{bmatrix}$$

■ ■

6.4.2 Classification of Buses

Having found the bus admittance matrix for a system from Equation 6.18, calculation of bus voltages requires values for the injected currents. These injected currents are almost never known directly. However, if certain other parameters are known, the currents can be calculated. Referring to Figure 6.5, the net real and reactive powers being injected into the system at bus i is

$$P_i + jQ_i = V_i I_i^* \qquad (6.19)$$

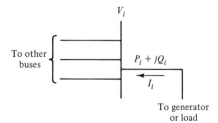

Figure 6.5 Calculation of injected currents.

6.4 POWER FLOW ANALYSIS

TABLE 6.2

	$|V|$	α	Net P	Net Q
Load bus	Calculated	Calculated	Known	Known
Voltage control bus	Known	Calculated	Known	Calculated
Swing bus	Known	Known	Calculated	Calculated

where I_i^* means the complex conjugate of current I_i. Net real and reactive powers being injected means the P and Q being generated at a bus minus the P and Q of any loads at that bus. Solving for I_i,

$$I_i = \frac{P_i - jQ_i}{V_i^*} \tag{6.20}$$

Equation 6.20 states that if the net real and reactive powers being injected into a bus are known, then the current being injected is a function of the voltage at that bus. These values of current as a function of voltage can be substituted into the left side of Equation 6.18. This substitution makes the entire equation a function of voltages only.

At buses where no generation is present, the demanded real and reactive powers of the load are known. These kinds of buses are known as load buses. At most buses where there are generators, the voltage magnitude and real power being generated are known. These buses are called voltage control buses. Since the reactive power generated is not known for a voltage control bus, an estimate of Q to be used in Equation 6.20 is needed. This procedure will be explained in detail later in this chapter. In some cases generator buses may be treated as load buses if a reactive power generation is specified at that bus instead of a voltage magnitude.

The real power being supplied by all the generators to the system cannot be fixed in advance of a system study because the losses in the system are not known until the analysis is complete. Therefore, one generator bus, called the swing bus, does not have a specified real-power generation. Instead, the swing bus has a specified voltage, magnitude, and angle, and all other bus voltage angles are referenced to it. The swing bus serves the function of supplying or demanding power to or from the system to keep the total system power generated and consumed in balance.

The three types of buses and the quantities known at each type of bus are given in Table 6.2. In Table 6.2 $|V|$ means the magnitude of the bus voltage, and α is the angle of the bus voltage referenced to the swing bus.

6.4.3 Load Bus Equations

The net injected real and reactive powers at a load bus are known quantities, as indicated by Table 6.2. Therefore, the current being injected at a load bus can be given as a function of the voltage at that bus by Equation 6.20. For the system shown in Figure 6.3, bus 3 is designated as a load bus. The nodal equation for node 3 can be written

$$\frac{P_3 - jQ_3}{V_3^*} = Y_{31}V_1 + Y_{32}V_2 + Y_{33}V_3 \tag{6.21}$$

All quantities in Equation 6.21 are known except the voltages. However, equation 6.21 is a nonlinear equation. All nodal equations in a power flow analysis are nonlinear. Solution of sets of nonlinear equations require special techniques.

6.4.4 Solution of Nonlinear Equations

To introduce a method of solution for nonlinear equations, a one-dimensional quadratic equation will be used:

$$f(x) = x^2 - 6x + 5 = 0 \tag{6.22}$$

This equation can be written in the form

$$x = \tfrac{1}{6}x^2 + \tfrac{5}{6} = F(x) \tag{6.23}$$

Figure 6.6 shows a graphic representation of the functions $f(x)$, $F(x)$, and the straight line $y = x$. The graph and the definition of $F(x)$ show that the values

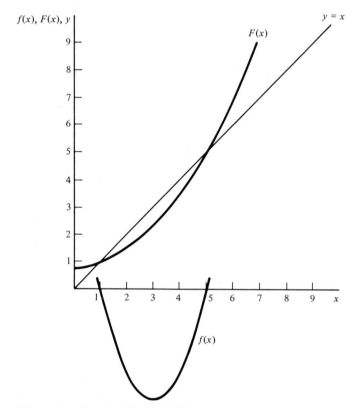

Figure 6.6 Graph of $f(x)$ and $F(x)$.

6.4 POWER FLOW ANALYSIS

of x when $x = F(x)$ are the roots of $f(x)$. Therefore, when the solution is found for $x = F(x)$, Equation 6.22 is also solved. Equation 6.23 can be solved by an iterative procedure. This method requires an initial estimate of the value of x. Let the initial estimate be indicated by $x^{(0)}$. Choose $x^{(0)} = 3$. The value of x is substituted into $F(x)$:

$$F(3) = \tfrac{1}{6}(3)^2 + \tfrac{5}{6} = 2.33 \tag{6.24}$$

Equation 6.24 indicates that the value of x after the first calculation, or iteration, is $x^{(1)} = 2.33$. This value of x is substituted into $F(x)$:

$$F(2.33) = \tfrac{1}{6}(2.33)^2 + \tfrac{5}{6} = 1.74 \tag{6.25}$$

The value of x after the second iteration is $x^{(2)} = 1.74$. The process is continued until two consecutive iterations of x are within some predefined tolerance, ε, of each other. Once the test shown in Equation 6.26 is satisfied, the iterative process is stopped:

$$\left| x^{(k)} - x^{(k+1)} \right| < \varepsilon \tag{6.26}$$

Equation 6.26 can also be written

$$\left| x^{(k)} - F(x^{(k)}) \right| < \varepsilon \tag{6.27}$$

or

$$x^{(k)} = F(x^{(k)}) \pm \varepsilon \tag{6.28}$$

The iterative procedure is illustrated in Figure 6.7.

This method is called the Gauss iterative method. A disadvantage of this method is that as the solution is approached the iterative steps become smaller. Therefore, a large number of iterations will probably be needed for sufficient

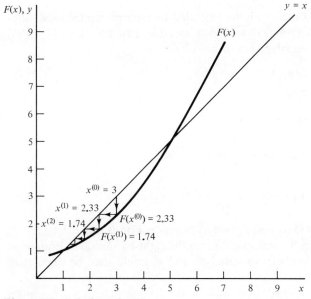

Figure 6.7 Solution of $F(x)$.

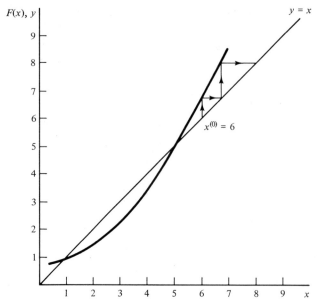
Figure 6.8 Divergence of Gauss method.

accuracy. Another disadvantage is illustrated in Figure 6.8. If the initial estimate is not a "good" estimate, the procedure will not converge toward the solution, but actually diverge away.

Other methods of solution which eliminate or reduce these disadvantages exist. For our purpose in this book, however, we shall use a modified version of the Gauss method.

The Gauss method can easily be extended to multiple dimensions. Suppose instead of one function $f(x)$ with one variable, two functions $f_1(x_1, x_2)$ and $f_2(x_1, x_2)$ with two variables are to be solved:

$$f_1(x_1, x_2) = x_1^2 - x_1 x_2 + \tfrac{3}{2} = 0 \qquad (6.29)$$

$$f_2(x_1, x_2) = x_2^2 + x_1 x_2 - 5 = 0 \qquad (6.30)$$

These functions can be written

$$x_1 = x_2 - \frac{3}{2x_1} = F_1(x_1, x_2) \qquad (6.31)$$

$$x_2 = -x_1 + \frac{5}{x_2} = F_2(x_1, x_2) \qquad (6.32)$$

Again an initial estimate of x_1 and x_2 would be made. For example, let $x_1^{(0)} = 1$ and $x_2^{(0)} = 2$. These values are substituted into the right-hand side of Equations 6.31 and 6.32. The first iterative values of x_1 and x_2 could then be calculated:

$$x_1^{(1)} = 2 - \frac{3}{2(1)} = 0.5 \qquad (6.33)$$

6.4 POWER FLOW ANALYSIS

$$x_2^{(1)} = -1 + \frac{5}{2} = 1.5 \tag{6.34}$$

The iterations would continue until both

$$\left|x_1^{(k+1)} - x_1^{(k)}\right| < \varepsilon \tag{6.35}$$

and

$$\left|x_2^{(k+1)} - x_2^{(k)}\right| < \varepsilon \tag{6.36}$$

were satisfied.

The speed of convergence of this method can be enhanced by a simple modification. As soon as a new value of a variable is available, it is used in all subsequent calculations. For instance, the first iterative values of x_1 and x_2 in the previous exercise would be calculated as follows:

$$x_1^{(1)} = 2 - \frac{3}{2(1)} = 0.5 \tag{6.37}$$

$$x_2^{(1)} = -0.5 + \frac{5}{2} = 2.0 \tag{6.38}$$

This method is known as the Gauss-Seidel method. In this book the Gauss-Seidel method will be used to solve nonlinear power flow equations, such as Equation 6.21.

Equation 6.21 can be written

$$V_3 = \frac{1}{Y_{33}}\left[\frac{P_3 - jQ_3}{V_3^*} - (Y_{31}V_1 + Y_{32}V_2)\right] \tag{6.39}$$

Written in this form, the equation for node 3 and all other nodal equations can be solved by direct application of the Gauss-Seidel method.

6.4.5 Voltage Control Buses

At a voltage control bus the net real power injected into the system and the voltage magnitude are specified. Bus 2 of Figure 6.3 is designated a voltage control bus. A nodal equation similar to Equation 6.39 can be written for bus 2:

$$V_2 = \frac{1}{Y_{22}}\left[\frac{P_2 - jQ_2}{V_2^*} - (Y_{21}V_1 + Y_{23}V_3)\right] \tag{6.40}$$

Equation 6.40 requires not only estimates for the voltages at all three buses, but also a value of net injected reactive power at bus 2 which is not specified. However, Q_2 can be estimated. From Equation 6.19 it is clear that

$$V_2 I_2^* = P_2 + jQ_2 \tag{6.41}$$

An expression for I_2 can be obtained from Equation 6.18. Making this substitution, Equation 6.41 becomes

$$V_2(Y_{21}V_1 + Y_{22}V_2 + Y_{23}V_3)^* = P_2 + jQ_2 \tag{6.42}$$

With the latest estimated values of voltages, Q_2 can be estimated as

$$Q_2 = \text{Im}[V_2(Y_{21}V_1 + Y_{22}V_2 + Y_{23}V_3)^*] \tag{6.43}$$

where Im means "the imaginary part of."

With Q_2 estimated from Equation 6.43, a new value for V_2 can be obtained from Equation 6.40. However, voltage V_2 has a specified magnitude. Only the angle calculated from Equation 6.40 is used. The new updated value of V_2 would have its given specified magnitude with an angle calculated from Equation 6.40.

In some instances a maximum and minimum value may be set on the net injected reactive power at a voltage control bus. If the Q estimated by an equation like 6.43 violates one of these constraints, then Q would be set to the maximum or minimum value, and the bus would be treated as a load bus.

6.4.6 The Swing Bus

The magnitude and angle of the voltage are specified at the swing bus. Therefore, the equation corresponding to the swing bus can be eliminated from Equations 6.18. So, for an N bus system, only $N - 1$ equations must be solved to determine all the system voltages. The net real and reactive powers being injected at the swing bus can be determined after the other system voltages have been calculated. The sum of real and reactive powers leaving the swing bus on all system elements connected to it, other than generators and loads, is the net injected real and reactive powers at the swing bus. For example, if bus 1 of Figure 6.3 is designated the swing bus, P_1 and Q_1 could be calculated after V_2 and V_3 have been determined by procedures described earlier:

$$P_1 + jQ_1 = V_1(Y_{11}V_1 + Y_{12}V_2 + Y_{13}V_3)^* \tag{6.44}$$

After this calculation, the injected real and reactive powers and the voltage magnitudes and angles are known at every bus in the system. The real- and reactive-power flows can then be found on any element in the system. For instance, the complex power flow from bus 2 to bus 3 on the element connecting buses 2 and 3 of the system in Figures 6.3 and 6.4 is

$$P_{23} + jQ_{23} = V_2[(V_2 - V_3)Y_e]^* \tag{6.45}$$

6.4.7 Initial Voltage Estimates

As we have seen, in order to solve the nonlinear equations for voltages of load and voltage control buses, initial estimates of these voltages must be made. Since the nominal operating voltages of a system are usually chosen as per unit bases, a good estimate of the magnitudes of the load bus voltages is 1.0 pu. The voltage angles throughout a system usually only vary a few degrees. By setting the swing bus voltage angle to zero, the other bus voltage angles will probably be near zero also. Therefore, 0° is a natural first approximation for all voltage angles.

6.4 POWER FLOW ANALYSIS

EXAMPLE 6.9

For the system shown below and the data given, find the voltages V_1, V_2, and V_3 after one iteration of a Gauss-Seidel load flow:

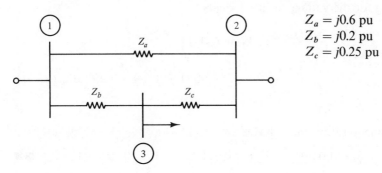

$Z_a = j0.6$ pu
$Z_b = j0.2$ pu
$Z_c = j0.25$ pu

Bus	Type	Voltage magnitude	Voltage angle	Net P	Net Q
1	Swing	1.04 pu	0°	—	—
2	Voltage control	1.02 pu	—	1.2 pu	—
3	Load	—	—	−1.5 pu	−0.5 pu

The impedances of the elements must be represented as admittances:

$$Y_a = -j1.67 \text{ pu} \quad Y_b = -j5.0 \text{ pu} \quad Y_c = -j4.0 \text{ pu}$$

Now the bus admittance matrix can be assembled:

$$\begin{bmatrix} -j6.67 & j1.67 & j5.0 \\ j1.67 & -j5.67 & j4.0 \\ j5.0 & j4.0 & -j9.0 \end{bmatrix}$$

The initial voltage estimates are

$$V_1 = 1.04\underline{/0°} \quad V_2 = 1.02\underline{/0°} \quad V_3 = 1.0\underline{/0°}$$

Since no equation must be solved for the swing bus, we start with bus 2:

$$V_2 = \frac{1}{Y_{22}} \left[\frac{P_2 - jQ_2}{V_2^*} - (Y_{21}V_1 + Y_{23}V_3) \right]$$

Since bus 2 is a voltage control bus, Q_2 must be calculated first:

$$\begin{aligned} Q_2 &= \text{Im}[V_2(Y_{21}V_1 + Y_{22}V_2 + Y_{23}V_3)^*] \\ &= \text{Im}\{1.02\underline{/0°}[1.67\underline{/90°}(1.04\underline{/0°}) + 5.67\underline{/-90°}(1.02\underline{/0°}) \\ &\quad + 4.0\underline{/90°}(1.0\underline{/0°})]^*\} \\ &= 0.048 \end{aligned}$$

Solving for V_2 leads to

$$V_2 = \frac{1}{5.67\underline{/-90°}} \left\{ \frac{1.2 - j0.048}{1.02\underline{/0°}} - [1.67\underline{/90°}(1.04\underline{/0°}) + 4.0\underline{/90°}(1.0\underline{/0°})] \right\}$$
$$= 1.04\underline{/11.5°}$$

Since bus 2 is a voltage control bus, we use the angle only:

$$V_2 = 1.02/\underline{11.5°}$$

Solving for the voltage at bus 3 yields

$$V_3 = \frac{1}{Y_{33}}\left[\frac{P_{33} - jQ_3}{V_3^*} - (Y_{31}V_1 + Y_{32}V_2)\right]$$

$$= \frac{1}{9.0/\underline{-90°}}\left\{\frac{-1.5 + j0.5}{1.0/\underline{0°}} - [5.0/\underline{90°}(1.04/\underline{0°}) + 4.0/\underline{90°}(1.02/\underline{11.5°})]\right\}$$

$$= 0.969/\underline{-4.5°}$$

After one iteration the voltages are

$$V_1 = 1.04/\underline{0°} \qquad V_2 = 1.02/\underline{11.5°} \qquad V_3 = 0.969/\underline{-4.5°} \qquad ■■$$

6.4.8 Results Obtained from a Load Flow Analysis

Load flow studies are used in planning the expansion of existing power systems. For a forecasted future load and generation schedule, the voltage magnitude and angle at each bus and the real and reactive powers flowing on each line can be determined. This information will indicate possible problems, such as low voltages or overloaded lines. Necessary corrections, such as new construction or upgrading equipment, can then be initiated before the system experiences the forecasted conditions.

Load flow studies are also used in contingency studies. This means that given a particular system studies are performed to determine if problems will result if one or more lines or generators are removed from service. From these results the reliability and security of the system can be determined.

6.5 CONTROL OF POWER FLOW

If unfavorable conditions are detected from a power flow analysis, some corrective action should be taken. This action could consist of construction of new transmission lines or generating stations to alleviate overloading in part of the system. However, many solutions to possible problems are much simpler and less expensive. Quite often the answer is to change the operating conditions of the system. That is, change where the real and reactive powers are generated and alter their path from the point of generation to the point of consumption. The discussion on the control of power flow will start at the synchronous generator.

6.5.1 Synchronous Generators

The generator to be discussed will be part of a large power system with several other generating stations. The system is so large and "strong" that every

6.5 CONTROL OF POWER FLOW

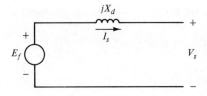

Figure 6.9 Equivalent circuit of a synchronous generator.

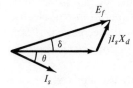

Figure 6.10 Phasor diagram of overexcited generator.

bus appears to be an infinite bus. That is, the voltage at the generator bus will not be altered by changes in the generator's operating conditions.

The single-phase equivalent circuit of a synchronous generator is shown in Figure 6.9.

Analysis in Chapter 3 showed that for a generator operating at a constant speed, the voltage E_f is a function of the dc excitation on the rotor. The phasor diagram for a generator supplying reactive power to the system (overexcited) is shown in Figure 6.10. If the real-power output is held constant while the dc excitation on the rotor is varied, some interesting things happen. Since the real-power output is constant, the projection of I_s onto V_s must remain constant. Therefore, if the magnitude of E_f with the rotor dc excitation is changed, the magnitude of I_s and the angles θ and δ also change. This event is illustrated in Figure 6.11. Figure 6.11 shows that the generation of reactive power can be

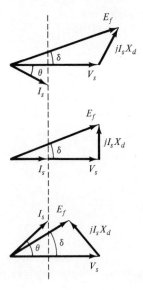

Figure 6.11 Varying E_f while maintaining constant power output.

controlled by varying E_f through the rotor dc excitation while maintaining a constant real-power output.

In Chapter 3 real power out of a generator was given by

$$P = \frac{V_s E_f}{X_d} \sin \delta \qquad (6.46)$$

Equation 6.46 indicates that if V_s and E_f are held constant, the real-power output is directly dependent on δ. Of course, in steady-state operation real electric power output can only be increased by an increase of real mechanical power input. Therefore, δ must also be directly dependent on the mechanical power input.

In Chapter 3 an equation for reactive-power output was derived:

$$Q = \frac{-V_s}{X_d}(V_s \cos \delta - E_f) \qquad (6.47)$$

If V_s and E_f of Equation 6.47 are held constant, Q would appear to be affected by changes in δ or changes in mechanical power input. However, the normal operating range for δ is less than 15°. For small changes of δ in this operating range, the cosine term is relatively unchanged while the sine term will vary significantly.

From this discussion the reactive power out of a generator appears to be controlled by adjusting the rotor dc excitation, and the real-power output appears to be controlled by varying the mechanical power input. By controlling the amount of real and reactive power generated at each generating station in a system, the general flow of power in the system can be controlled.

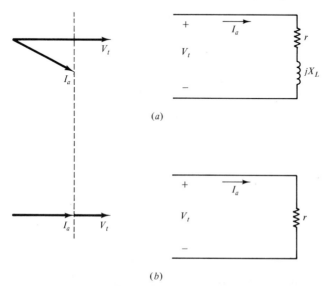

Figure 6.12 Current needed to supply (a) an inductive load and (b) a purely resistive load.

6.5 CONTROL OF POWER FLOW

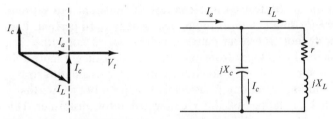

Figure 6.13 Capacitor bank at an inductive load.

6.5.2 Capacitor Banks

Most loads in a power system are inductive. Reactive power must, therefore, be generated to supply this inductive load. Only the real power that is delivered to the load actually does real mechanical work. Figure 6.12 illustrates that delivering the same amount of real power to a load with a reactive part requires more current than supplying a load with no reactive part. Larger currents mean greater voltage drops along transmission and distribution lines and more $I^2 r$ line losses. Both of these conditions are undesirable.

One way to overcome these disadvantages is to generate reactive power at the location that it is needed. Since capacitors consume reactive power, which is the same as supplying vars, the installation of a capacitor bank at the load bus will have this effect. This phenomenon is illustrated in Figure 6.13. The result of adding the properly sized capacitor bank at the load is that no reactive power must flow through the lines of the system to supply the load reactance.

6.5.3 Tap-Changing Transformers

In Chapter 5 the flow of real and reactive power from one terminal of a transmission line toward the other terminal was shown to be dominated by the terms

$$P_{12} = V_1 V_2 b(\theta_1 - \theta_2) \tag{6.48}$$

$$Q_{12} = V_1 b(V_1 - V_2) \tag{6.49}$$

If Equations 6.48 and 6.49 are expressed in per unit with voltage bases chosen as nominal voltage in each part of the system, voltage V_1 and V_2 will be approximately equal to 1.0 and Equations 6.48 and 6.49 become

$$P_{12} = b(\theta_1 - \theta_2) \tag{6.50}$$

$$Q_{12} = b(V_1 - V_2) \tag{6.51}$$

These equations clearly indicate that the variables that control real- and reactive-power flows on a line are voltage angles and voltage magnitudes, respectively. These variables can be controlled by the use of tap changing under load transformers (TCUL) or regulating transformers.

Such transformers change the number of turns used in one of the coils by changing the location of the tap on the coil of the transformer. Therefore, the ratio of transformation of the transformer changes for different tap settings. This operation allows for small changes in voltage, usually ±10 percent. Regulating transformers do not transform between voltage levels but rather just make small adjustments in voltages from one side to the other. Figure 6.14 shows an illustration for a voltage magnitude-controlling regulating transformer and a phase-shifting regulating transformer. Figure 6.14a shows that a voltage in phase with V_{an} is tapped off the Y-connected autotransformer. This

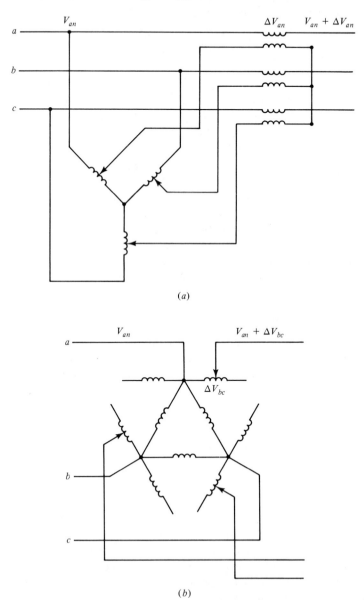

Figure 6.14 Regulating transformers: (*a*) magnitude and (*b*) phase.

6.5 CONTROL OF POWER FLOW

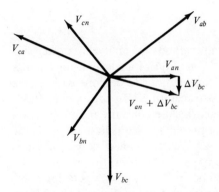

Figure 6.15 Phasor diagram for phase-shifting transformer.

voltage is then summed to V_{an} through the transformer in the line of phase a. Since all voltages in each phase are in phase with each other, only the magnitude of V_{an} changes. In Figure 6.14b the coils at each point of the Δ are wound on a core with the Δ phase coil they parallel. If the voltages of Figure 6.14b are in abc sequence, the corresponding phasor diagram is shown in Figure 6.15. As long as ΔV_{bc} is small, the magnitude of V_{an} will remain relatively unchanged while its angle is shifted.

With these types of transformers voltage magnitudes and angles can be adjusted to help control the flow of real and reactive power on transmission and distribution lines. For instance, if two parallel lines are supplying a reactive load at a bus and line 1 has a voltage magnitude-regulating transformer located at the load bus, the flow of reactive power in the lines can be controlled. This circuit is shown in Figure 6.16. If line 2 is carrying too much of the Q load, the regulating transformer in line 1 could be set to raise the voltage at the load bus. If the load bus voltage is raised, then the voltage drop across line 2 must be less. From Equation 6.49 we see that under these conditions the reactive-power flow of line 2 would be decreased. The var flow on line 1 would, therefore, have to increase. This additional var flow causes the voltage on the line side of the transformer to be low. However, the transformer steps this voltage up to the desired level at the load bus. In this manner some of the flow of reactive power to the load bus has been shifted from line 2 to line 1.

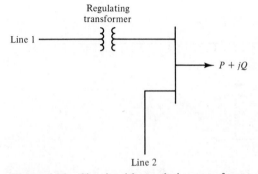

Figure 6.16 Circuit with regulating transformer.

6.6 SUMMARY

This chapter has presented useful tools and methods for analyzing power systems under normal operating conditions. A method of analysis using the per unit system has been discussed that allows calculations to be done without the complicated transferal of impedances across transformers. In addition, methods of solving the nonlinear equations presented by the constant P and Q representation of loads on a power system were presented. With these methods the voltage magnitude, voltage angle, and real- and reactive-power flow at all buses in a power system can be calculated for any particular operating point.

Since the operating conditions of the system for any given operating point can be found, methods of controlling voltages and power flows were studied next. Methods of using generator excitation and mechanical input, capacitor banks, and transformers were examined for this purpose.

6.7 PROBLEMS

6.1. A 150-kVA, 2300/230-V, 60-Hz, single-phase transformer has the following series impedances:

$$r_1 = 0.22 \, \Omega \quad r_2 = 0.002 \, \Omega$$
$$X_1 = 2.0 \, \Omega \quad X_2 = 0.022 \, \Omega$$

Using the transformer ratings as base, calculate the per unit series impedance of the transformer from the high-voltage side and the low-voltage side.

6.2. Draw an impedance diagram of the three-phase system in Figure 6.17 showing all impedances in per unit. The bases of the $j50 \, \Omega$ transmission line are 100 MVA and 140 kV. The manufacturer's data for each device is given as follows:

Generator 1:	12 kV	50 MVA	$X = 8\%$
Generator 2:	12 kV	40 MVA	$X = 8\%$
T_1: three-phase	13/138 kV	45 MVA	$X = 10\%$
T_2: three, single-phase	6.9/39.8 kV	25 MVA	$X = 8\%$
T_3: three-phase	69/138 kV	75 MVA	$X = 12\%$
T_4: three-phase	12/138 kV	80 MVA	$X = 12\%$

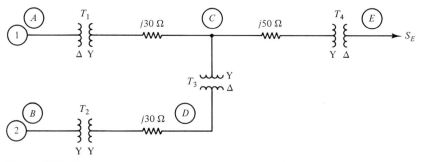

Figure 6.17 System.

6.7 PROBLEMS

6.3. Repeat Problem 6.2 with the following changes in transformer banks:

T_1: three, single-phase 13/80 kV 15 MVA $X = 9\%$
T_3: three, single-phase 69/80 kV 25 MVA $X = 13\%$

6.4. With the bases chosen and the data given in Problem 6.2, calculate the magnitude of the voltage at the terminals of each generator of the system in Figure 6.17 when the load in section E is 60 MW at a power factor of 0.9 lag and the voltage at the load is 11 kV. Assume that in section C the current from generator 1 is equal to the current from generator 2.

6.5. Figure 6.18 shows a power system and its impedances in per unit. Assemble the bus admittance matrix for this system.

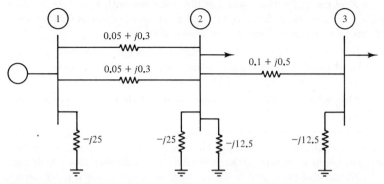

Figure 6.18 System.

6.6. The impedances of a power system are shown in Figure 6.19. Determine the injected currents at each bus if $V_1 = 1.03\underline{/0°}$, $V_2 = 1.02\underline{/-1°}$, $V_3 = 1.0\underline{/-3°}$, and $V_4 = 1.01\underline{/-3°}$.

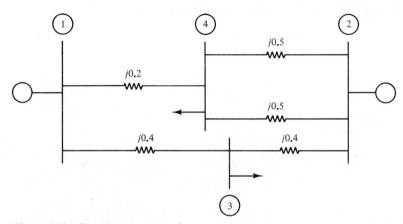

Figure 6.19 System.

6.7. Using the Gauss-Seidel method, perform two iterations in the solution of the following set of equations:

$$4X_1^2 + 3X_1X_3 + X_1X_3 = 35$$
$$X_1X_2 + 2X_2^2 + X_2X_3 = 18$$
$$4X_1X_3 + 3X_2X_3 + 7X_3^2 = 62$$

Use initial estimates of $X_1 = X_2 = X_3 = 2$.

6.8. Perform one iteration of a Gauss-Seidel power flow solution of the system of Problem 6.5. Buses 2 and 3 are load buses with per unit loads of $0.25 + j0.12$ pu and $0.3 + j0.15$ pu, respectively. Bus 1 is the swing bus with $V_1 = 1.04\underline{/0°}$. Perform the bus voltage calculations in numerical order.

6.9. Perform one iteration of a Gauss-Seidel power flow solution of the system of Problem 6.6. Buses 3 and 4 are load buses with per unit loads of $0.23 + j0.19$ pu and $0.42 + j0.13$ pu, respectively. Bus 1 is the swing bus with $V_1 = 1.03\underline{/0°}$. Bus 2 is a voltage control bus with $|V_2| = 1.02$ and a generated power of 0.24 pu. Perform the bus voltage calculations in numerical order.

6.10. For the system shown in Figure 6.19 determine the complex power flow from the lower-numbered bus toward the higher-numbered bus of each transmission line if $V_1 = 1.03\underline{/0°}$, $V_2 = 1.02\underline{/-1°}$, $V_3 = 1.0\underline{/-3°}$, and $V_4 = 1.01\underline{/-3°}$.

6.11. Repeat Problem 6.10 using the approximations for real- and reactive-power flows given in Equations 6.50 and 6.51.

6.12. A transformer supplies a three-phase load through a transmission line which has a per phase impedance of $0.1 + j0.3$ pu. The load is $0.5 + j0.3$ pu and is supplied at a voltage of 1.0 pu at the load. If the voltage at the transformer does not change, what per unit value of capacitive reactance must be added at the transformer to make the power factor of the load on the transformer 0.9 lag?

6.13. Using a phase-shifting transformer similar to the one described in Section 6.5.3, determine how much the phase of a voltage can be shifted and still hold the change in magnitude of the voltage to 1 percent.

6.14. Using the same assumptions made to derive the approximate real- and reactive-power flows along a line (Equations 6.50 and 6.51), show that for a generator if V_s is constant then the magnitude of E_f controls the amount of reactive power generated and δ controls the amount of real power generated.

Chapter 7

Electric Motors

7.1 INTRODUCTION

The remaining portion to study of a power system is its loads. Figure 7.1 shows that the loads are connected to the transmission system, usually at distribution level voltages. Most of the loads (50 to 70 percent) are electric motors, with the remaining portion consisting primarily of heating and lighting loads. This chapter will consider the electric motor portion, which can be categorized into three types: synchronous, induction, and dc motors.

7.2 ELEMENTARY MOTOR OPERATION

Consider again the elementary machine examined in Chapter 3 and shown in Figure 7.2. An N-turn coil carrying i amperes of current is suspended in a magnetic field with flux density B. Experiments have shown that the sides of the

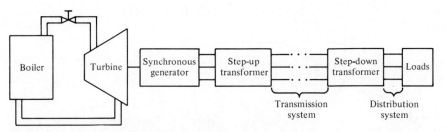

Figure 7.1 Electric loads within a power system.

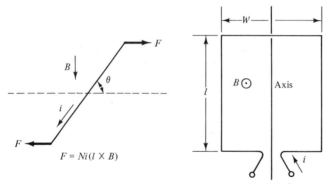

Figure 7.2 Elementary machine. (a) Top view and (b) front view.

coil marked with length l experience a force of

$$F = Ni(l \times B) \quad \text{N} \tag{7.1}$$

The direction of vector l is the direction of current flow, and the cross product of l and B yields a force with direction as shown in Figure 7.2a. If the coil is allowed to move, the tangential component of force F will cause rotation of the coil about its axis. The magnitude of the tangential force is derived from Figure 7.3.

$$\begin{aligned} F_T &= F \sin \theta \\ &= NilB \sin \theta \end{aligned} \tag{7.2}$$

Torque exerted on the coil by both tangential forces is easily derived from the force multiplied by the moment arm and results in

$$\tau = NilBW \sin \theta \quad \text{N·m} \tag{7.3}$$

Recall from Equation 2.32 that Ni is called MMF and is designated F_r for the coil in Figure 7.2. Therefore, Equation 7.3 can be rewritten

$$\tau = F_r BlW \sin \theta \quad \text{N·m} \tag{7.4}$$

Figure 7.3 shows that when $\theta = 0°$, the torque on the coil is 0, and F_r is aligned with B. Also, at any angle $\theta \neq 0°$ torque is developed such that, in the presence of damping on the coil shaft, the coil will rotate to and stop at the position where F_r is aligned with B. This phenomenon of torque developing to align a

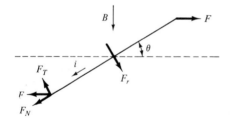

Figure 7.3 Tangential and normal components of force F.

7.2 ELEMENTARY MOTOR OPERATION

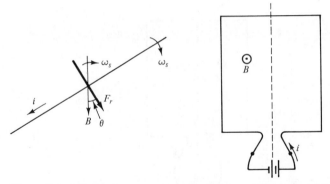

Figure 7.4 Elementary motor operated as a synchronous motor.

coil F_r suspended within a magnetic field is the effect used by synchronous, induction, and dc motors to develop rotation. However, two significant problems with this kind of operation are illustrated in Equation 7.4:

1. The coil cannot sustain rotation. This effect occurs because at $\theta = 0°$ no torque is applied to the coil.
2. The developed torque is not constant. This effect occurs because the torque varies with $\sin \theta$. Variation in torque is a source of vibration and noise. Both of these effects are undesirable for motor operation.

Motors are designed to overcome these two problems, and the way they do so defines the difference between synchronous, induction, and dc motors.

For synchronous motor operation rotation is sustained by applying direct current to the coil and rotating B about the coil axis. Figure 7.4 shows the elementary motor operated in this fashion. As B rotates at ω_s (called the synchronous speed), the resulting torque causes the coil and therefore F_r to be dragged along behind at an angle θ. Angle θ settles into a constant value such that the torque developed on the coil matches the torque required by the mechanical load attached to the coil shaft. Thus, the coil also rotates at synchronous speed, and variation of the torque is minimized because θ is constant.

At standstill θ will vary from 0° to 360°, and the average torque will be 0. Therefore, a synchronous motor has no starting torque.

For induction motor operation rotation of the coil is also sustained by rotating B. However, instead of applying direct current to the coil, the coil terminals are shorted together. Figure 7.5 shows the elementary motor operated in this fashion. As B rotates, the changing flux linkages of the coil induces Faraday law voltage on the coil. This voltage causes current i to flow in the coil, thus establishing F_r. B and F_r rotate at synchronous speed. The coil, on the other hand, rotates at a speed of ω_r such that $\omega_r < \omega_s$. Speed ω_r must be strictly less than ω_s in order for B to vary with time as it links the coil. This condition settles down to a steady-state operating point where θ and ω_r remain constant. Then the torque developed on the coil matches the torque required

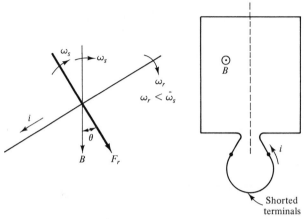

Figure 7.5 Elementary motor operated as an induction motor.

by the mechanical load attached to the shaft. Thus, the coil sustains rotation and the torque variation is minimized because θ is constant.

For dc motor operation rotation is sustained by holding B and F_r stationary with respect to each other. Figure 7.6 shows the elementary motor operated in this fashion. The current in the coil is direct current, and ideally angle θ between B and F_r should be kept at 90° in order to develop maximum torque. However, this condition cannot occur for the motor shown in Figure 7.6. F_r will rotate such that $\theta = 0°$. Since B does not rotate, sustaining rotation by the synchronous or induction motor method cannot be achieved with a dc motor. Instead, when F_r is near the positions of $\theta = 0°$, the direction of current flow in the coil is reversed. Figure 7.7 shows this operation. As θ reaches 0°, the direction of current i is reversed. If the rotational inertia of the coil shaft load carries the coil past $\theta = 0°$, the forces that develop on the coil are as shown in Figure 7.7b. Note that the tangential force defined by Equation 7.2 is oriented such that the torque developed on the coil has a constant direction—clockwise in this case. However, the torque magnitude (at least for this elementary motor) varies with θ. Since this effect is not desirable, dc motors are designed with coils

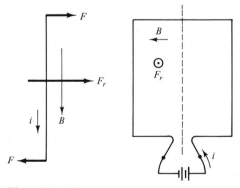

Figure 7.6 Elementary motor operated as a dc motor.

7.3 SYNCHRONOUS MOTORS

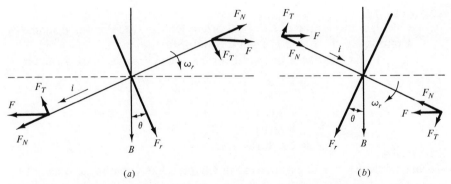

Figure 7.7 Effect of switching coil current direction on coil forces. (*a*) Current flow right to left across top of coil and (*b*) current flow left to right across top of coil.

on the rotating piece that are positioned to smooth out the torque and that do not rely on inertia to sustain rotation.

Each of the three motors described above are examined in more detail in the following sections. Bear in mind while reading these sections that the differences in design are prompted by (1) sustaining rotation and (2) providing constant torque during steady-state operation.

7.3 SYNCHRONOUS MOTORS

The design of synchronous motors is identical to the design of synchronous generators as discussed in Chapter 3. Recall that a synchronous machine has a structure as shown in Figure 3.9. The magnetic field that develops across the air gap between its stator and rotor must be made to rotate in order to sustain rotation. This motion is achieved by applying three-phase current excitation to the spatially distributed stator coils. This kind of excitation develops a rotating stator MMF, called F_s, as was illustrated in Example 3.5. If the rotor MMF, called F_r, is also rotating at the same speed as F_s, torque will be developed on the rotor.

Calculation of synchronous motor torque follows the same procedure as was used to calculate synchronous generator torque. Exactly the same result as Equation 3.33 is obtained; it is repeated here:

$$\tau^e = \frac{-3L_{sr}I_{max}I_r}{2} \sin \gamma \tag{7.5}$$

The difference between applying this equation to the synchronous generator and applying it to a synchronous motor is in the value of γ. For generator operation, γ was the angle between F_s and F_r measured from F_s to F_r. This angle was positive because F_s lagged F_r. However, for motor operation, F_r lags F_s and γ is negative. This condition results in τ^e having positive values for motor operation and negative values for generator operation.

EXAMPLE 7.1

Calculate the torque of electrical origin for the synchronous machine of Example 3.5 if γ is -0.77π rad for motor operation.

From Chapter 3 the pertinent machine parameters are as follows:

$$r = 0.5 \text{ m} \qquad i_a = 100 \cos 377t \text{ A}$$
$$l = 2.0 \text{ m} \qquad i_r = 8 \text{ A}$$
$$g = 0.005 \text{ m} \qquad \theta = 377t - 0.77\pi \text{ rad}$$
$$N_s = 4 \text{ turns} \qquad L_l = 2.65 \times 10^{-4} \text{ H}$$
$$N_r = 200 \text{ turns}$$

Torque of electric origin was derived in Chapter 3 at Equation 3.33 as

$$\tau^e = \frac{-3L_{sr}I_{max}I_r}{2} \sin \gamma$$

where

$$L_{sr} = \frac{4\mu_0 N_s N_r rl}{\pi g}$$

For the conditions of motor operation τ^e appears as

$$\tau^e = \frac{(-3)(4)(4\pi \times 10^{-7})(4)(200)(0.5)(2.0)(100)(8)\sin(-0.77\pi)}{(2)(\pi)(0.005)}$$

$$= 203.2 \text{ N}\cdot\text{m}$$

Derivation of an electric circuit model of a synchronous motor is performed in exactly the same manner as was done for the synchronous generator and results in the same model. The difference between applying the model to the motor as opposed to the generator is again a function of angle γ. For motor operation γ is negative because F_r lags F_s. Example 7.2 illustrates this difference. ∎∎

EXAMPLE 7.2

Calculate X_m, X_s, E_f, and E_s, and V_s for the motor of Example 7.1. The circuit model is shown below.

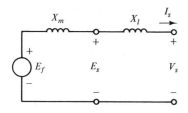

From Equation 3.41 and the expression for L_s in Equation 3.26,

$$X_m = 3\omega_s L_s = 3\omega_s \frac{2\mu_0 N_s^2 rl}{\pi g}$$

For the data of this machine

$$X_m = 2.895 \qquad \text{(see Example 3.11)}$$

7.3 SYNCHRONOUS MOTORS

Since
$$X_d = X_m + X_l$$
the data for this machine yields
$$X_d = 2.995 \quad \text{(see Example 3.11)}$$

Using Equation 3.40 and L_{sr} from Equation 3.26, the value of v_f is

$$v_f = \omega_s \frac{4\mu_0 N_s N_r r l}{\pi g} I_r \sin(\omega_s t + \gamma)$$

$$= \frac{(377)(4)(4\pi \times 10^{-7})(4)(200)(0.5)(2.0)(8)\sin(377t - 0.77\pi)}{(\pi)(0.005)}$$

$$= 772.1 \sin(377t - 0.77\pi)$$
$$= 772.1 \cos(377t - 1.27\pi) \text{ V}$$

Using Equations 3.40 and 3.41 and the phasor form of v_f, E_s is calculated in phasor form as

$$E_s = E_f - jX_m I_s$$
$$E_s = 546.0\underline{/131.4°} - j(2.895)(70.7\underline{/0°})$$
$$= 415.1\underline{/150.4°} \text{ V}$$

Solve for V_s from

$$V_s = E_f - jX_d I_s$$
$$= 546.0\underline{/131.4°} - j(2.995)(70.7\underline{/0°})$$
$$= 411.7\underline{/151.3°} \text{ V}$$

A vector diagram of these quantities appears as shown:

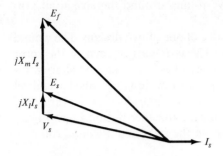

EXAMPLE 7.3

Calculate the three-phase power flowing out of the machine of Example 7.1 using the equation for P as a function of δ equation and the equation $S = VI^*$.

The former equation yields

$$P = \frac{3|E_f||V_s|}{X_d} \sin \delta$$

where $|E_f|$ and $|V_s|$ are rms quantities and δ is measured from E_f to V_s, or

$$\delta = \underline{/E_f} - \underline{/V_s}$$
$$= 131.4° - 151.3°$$
$$= -19.9°$$

Thus, P is calculated as

$$P = \frac{3 \times 546 \times 411.7 \times \sin(-20.0°)}{2.995}$$
$$= -77{,}000 \text{ W}$$

From the definition of complex power

$$P = \text{Re } S = \text{Re } 3V_s I_s^*$$
$$= 3V_s I_s \cos(\underline{/V_s} - \underline{/I_s})$$
$$= 3(411.7)(70.71) \cos(151.3° - 0°)$$
$$= -77{,}000 \text{ W}$$

The negative values of power flow out of the motor are expected. Since the motor is converting electric energy into mechanical energy, real power must flow into the electric terminals of the machine. ■■

7.4 THREE-PHASE INDUCTION MOTORS

The stator of a three-phase induction motor is essentially identical to the stator of a three-phase synchronous machine. When it is excited by a balanced source, an MMF wave is developed that at any point in time is sinusoidally distributed around the air gap. This wave rotates around the air gap at synchronous speed.

The rotor of an induction motor can be of one of two designs. The wound rotor has windings similar to the stator. The terminals of each of the rotor windings are connected to insulated slip rings. These slip rings are shorted together in normal operation. The second type of rotor design is called a squirrel cage rotor. It consists of copper or aluminum bars embedded in the surface of the rotor. The ends of all these bars are shorted together by end rings. Figure 7.8 shows a sketch of a squirrel cage rotor on the right and a wound rotor on the left.

7.4.1 Operation of the Three-Phase Induction Motor

The stator magnetic field of a three-phase induction motor is very similar to that of a synchronous machine. Since in normal operation the rotor coils are short-circuited, all of the excitation for the induction motor must come from the stator. Voltages are induced on the rotor, and as a result current flows in the rotor. The rotor current creates an MMF wave that is sinusoidally distrib-

7.4 THREE-PHASE INDUCTION MOTORS

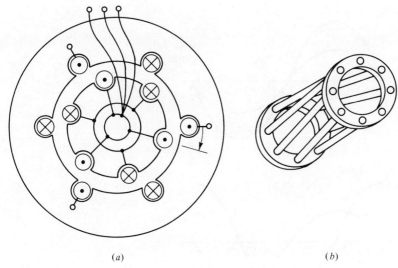

Figure 7.8 Wound rotor and squirrel cage rotor.

uted around the air gap and, as will be shown, travels at synchronous speed. Since both the stator and rotor MMF waves are traveling at synchronous speed and have constant amplitudes, their summation gives a constant magnitude, sinusoidally distributed resultant MMF wave that rotates at synchronous speed. This resultant MMF wave creates the flux that links the stator and rotor coils. The flux wave, therefore, is also sinusoidally distributed and always travels around the air gap at synchronous speed.

7.4.2 Rotor Quantities at Standstill

Consider an induction motor at standstill. As the rotating flux wave travels around the air gap, the flux linking each coil of a wound rotor or each bar of a squirrel cage rotor will change. Therefore, a voltage will be induced on the coils or bars. Since the rotor circuits are short-circuited, the induced voltages will cause current to flow on the rotor.

The flux wave is sinusoidal. Therefore, one point on the flux wave can be associated with the location of the maximum induced voltage on the rotor. That is, the maximum voltage will be induced at every location on the rotor when that part of the flux wave passes over it. This phenomenon means that the voltage induced on the rotor takes the form of a voltage wave traveling on the rotor at synchronous speed. The rotor currents and rotor MMF must, therefore, also be in the form of waves traveling at synchronous speed. This concept is illustrated in Figure 7.9.

Figure 7.9 shows the rotor of a squirrel cage induction motor not in its round form but as a straight line for easier illustration. At standstill the bars do not move. The flux density wave travels to the right at synchronous speed. This motion induces the voltage wave shown in Figure 7.9a. The voltage at each

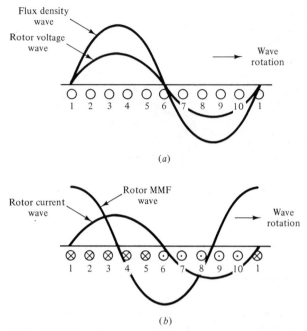

Figure 7.9 Traveling waves on rotor.

bar is equal to the magnitude of the voltage curve directly above or below the bar. Since the voltage wave passes over each bar at synchronous speed, the voltage on each bar will vary sinusoidally with time at synchronous frequency. At standstill the frequency of the rotor voltages and currents is synchronous frequency.

The current wave lags behind the voltage wave because of the rotor bars' leakage inductance. In Figure 7.9b the magnitude of the current in each bar corresponds to the magnitude of the current curve directly above or below the bar. The maximum of the rotor MMF wave is 90° behind the maximum of the current wave and is located at the magnetic axis of the currents flowing in the bars.

The interaction of the rotor currents and the flux density wave will result in a force on the rotor. As explained earlier, this force will act so as to try to align the rotor MMF wave with the flux density wave. Therefore, a three-phase induction motor has a torque applied at standstill. That is, a three-phase indication motor has a starting torque.

7.4.3 Running Rotor

The torque that develops at standstill causes the rotor to begin turning. The rotor will accelerate until it reaches a steady-state operating speed. With the rotor turning, the flux wave will no longer be traveling past the rotor at synchronous speed. The relative speed between the flux wave and the rotor is given

7.4 THREE-PHASE INDUCTION MOTORS

by

$$n_{rf} = n_s - n \tag{7.6}$$

where n_{rf} = relative speed between flux wave and rotor
n_s = relative speed between flux wave and stator, synchronous speed
n = mechanical speed of rotor

Since n_{rf} is the speed at which the flux wave passes over the rotor conductors, it is also the speed of the rotor voltage wave, the rotor current wave, and the rotor MMF wave with respect to the rotor. Following the same reasoning as with the rotor at standstill, the frequency of the rotor voltages and currents can be calculated as

$$\begin{aligned} f_r &= \frac{n_{rf}}{n_s} f \\ &= \frac{n_s - n}{n_s} f \end{aligned} \tag{7.7}$$

where f_r = frequency of rotor quantities
f = synchronous frequency of stator

Equation 7.7 can be rewritten

$$f_r = sf \tag{7.8}$$

where s is called the slip and is defined as

$$s = \frac{n_s - n}{n_s} \tag{7.9}$$

The value of slip varies from 1.0 at standstill to 0 at synchronous speed and is a measure of the relative speed between the rotor and flux wave.

As noted earlier, the speed of the rotor MMF with respect to the rotor is n_{rf}. If n_{rf} is added to the mechanical speed of the rotor n, the speed of the rotor MMF wave with respect to the stator is obtained:

$$n_{rf} + n = n_s - n + n = n_s \tag{7.10}$$

Equation 7.10 indicates that the rotor MMF wave travels at synchronous speed around the air gap at all normal operating slips.

EXAMPLE 7.4

A two-pole, 60-Hz, three-phase induction motor is operating at a slip of 3 percent. Find the frequency of the rotor voltages and currents and the speed of the rotor in rpm.

Synchronous speed is calculated for a machine with p poles as

$$\begin{aligned} n_s &= f\left(\frac{2}{p}\right) 60 \\ &= 60\left(\frac{2}{2}\right) 60 \\ &= 3600 \text{ rpm} \end{aligned}$$

From Equation 7.9 we have

$$s = \frac{n_s - n}{n_s}$$

$$n = n_s - sn_s = n_s(1-s)$$
$$= 3600(0.97)$$
$$= 3492 \text{ rpm}$$

Rotor frequency is calculated from Equation 7.8:

$$f_r = sf$$
$$= (0.03)60$$
$$= 1.8 \text{ Hz}$$

■■

If the flux wave is assumed to have constant magnitude, then the torque tending to move the rotor can be written as a function of the magnitude of the rotor MMF wave and its position with respect to the flux wave. The magnitude of the rotor MMF wave is directly proportional to the magnitude of the rotor current. Similarly, the position of the rotor MMF is a function of how much the rotor current wave lags the rotor voltage wave. Therefore, torque will increase as the magnitude of the rotor current increases and as the current lags less behind the voltage.

As the motor accelerates, the relative motion of the flux wave and the rotor decreases. Therefore, the rate of change of rotor flux linkages decreases, and the induced voltage and current decrease. This situation would tend to decrease torque. However, as the rotor speeds up, the frequency of the rotor quantities also decreases. This decrease means that the inductive leakage reactance of the rotor is decreased, and the current will not lag the voltage as much. Lower rotor frequency tends to increase torque. Therefore, the variation of torque as speed increases is not obvious.

For a typical motor the torque will increase with speed from standstill to a slip where maximum torque occurs. At speeds beyond that slip the torque will decrease. This concept is illustrated in Figure 7.10. As the speed of the rotor continues to increase, the relative speed between the flux wave and the rotor decreases. When the rotor reaches synchronous speed, no relative motion exists between it and the flux wave. Therefore, no voltage is induced on the rotor, and

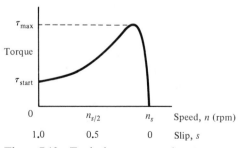

Figure 7.10 Typical torque-speed curve.

7.4 THREE-PHASE INDUCTION MOTORS

thus no current flows in the rotor coils. Without current flow, the rotor will not develop an MMF and there will be no torque on the rotor. Therefore, the normal operating point for an induction motor is between the slip where maximum torque occurs and a slip of zero. In this range all possible torques are available in a relatively small range of speeds.

If a motor is operating in the normal operating range of slip for a particular load, Figure 7.10 shows that if the load were increased, the motor would decrease in speed until the torque of the motor matched the required load torque. This action is limited by the value of maximum torque. If the load torque is raised higher than the value of maximum torque, the motor will slow down beyond the speed of maximum torque. At these speeds, according to Figure 7.10, less torque is produced, and the motor will stall. The maximum torque is sometimes called the breakdown torque.

7.4.4 Equivalent Circuit of the Induction Motor

In developing an equivalent circuit for a three-phase transformer bank, the bank was conveniently represented as being Y connected and analyzed on a single-phase basis. Equivalent Y connections with the same characteristics as the actual transformers were used to represent the devices that were not Y connected. The same method of analysis will be used for the induction motor. Therefore, all currents will be line values, and the voltages will be line-to-neutral values.

The flux wave, which is traveling around the air gap at synchronous speed, induces balanced voltages on the stator windings. These voltages are very similar to the voltage induced on the primary coils of a transformer by the flux in its core. The difference between the stator terminal voltage and the induced voltage in each phase is the voltage drop across its effective resistance and leakage reactance. This voltage drop also has a direct analog in the transformer equivalent circuit. The exciting current needed to create the air gap flux and supply the core losses must come from the stator since it is the only connection to a source. The stator equivalent circuit, so far, appears to be very similar to the primary of a transformer, and Figure 7.11 shows that it is.

In Figure 7.11 V_1 is the terminal voltage, E_1 is the induced voltage on the coil, r_1 is effective stator resistance, X_1 is stator leakage reactance, r_c represents

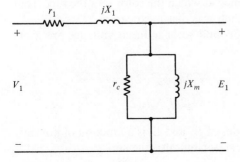

Figure 7.11 Stator circuit.

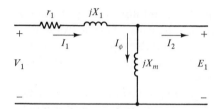

Figure 7.12 Simplified stator circuit.

core losses, and X_m is the magnetizing reactance. The magnetizing current in a transformer is usually in the range of 3 to 4 percent of rated full-load current. For an induction motor the magnetizing current can be 30 to 40 percent of full-load current. Consequently, the magnetizing reactance cannot be moved to the terminals of the motor, as in the transformer model, and still provide accurate modeling results. The core losses, however, are quite often assumed to be a constant value at all operating conditions and can be handled separately from the equivalent circuit analysis. Therefore, the stator equivalent circuit can be drawn as in Figure 7.12.

The currents I_2 and I_ϕ are the load component and exciting component of the stator current I_1, respectively.

The same flux wave that creates the voltage in the coils of the stator also induces a voltage on the rotor. Since both voltages are created by the same flux, a relationship exists between them. The voltage on the stator coils results from the time rate of change of the flux linkages in them:

$$e_1 = \frac{d\lambda}{dt} = \frac{dN_1 \phi_m \sin(\omega t + \alpha)}{dt} \qquad (7.11)$$

where N_1 is the effective number of turns in each stator coil. The flux term, $\phi_m \sin(\omega t + \alpha)$, represents the sinusoidally distributed flux wave with a maximum value of ϕ_m and initial displacement of α. The frequency of the flux wave is the synchronous frequency f of the stator.

The voltage generated on the rotor is calculated by the time rate of change of the flux linkages of the rotor coils:

$$e_r = \frac{d\lambda_2}{dt} = \frac{dN_2 \phi_m \sin(\omega_2 t + \alpha)}{dt} \qquad (7.12)$$

where N_2 represents the effective number of turns of the rotor circuit. The only difference in the flux term is the frequency at which the rotor sees the flux. That frequency is the rotor frequency f_r, which was discussed earlier.

The rms magnitudes of the stator- and rotor-induced voltages are given by Equations 7.13 and 7.14:

$$|E_1| = N_1 \omega \phi_m / \sqrt{2} = N_1 2\pi f_r \phi_m / \sqrt{2} \qquad (7.13)$$
$$|E_r| = N_2 \omega_2 \phi_m / \sqrt{2} = N_2 2\pi f_r \phi_m / \sqrt{2} \qquad (7.14)$$

The relationship between the magnitudes of E_1 and E_r is a function of not only the turns ratio between the rotor and stator but also the slip.

7.4 THREE-PHASE INDUCTION MOTORS

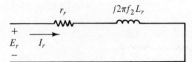

Figure 7.13 Rotor circuit.

$$\frac{|E_1|}{|E_r|} = \frac{N_1 f}{N_2 f_r} = \frac{N_1 f}{N_2 sf} = \frac{N_1}{N_2 s} \quad (7.15)$$

Since the rotor circuits are normally shorted, a resistance and leakage reactance are the only impedances that are represented in the rotor equivalent circuit. The equivalent circuit of one rotor circuit is shown in Figure 7.13.

Consider the leakage reactance of the rotor circuit. It is obviously frequency dependent. However, if sf is substituted for f_r, the leakage reactance becomes

$$2\pi f_r L_r = 2\pi f s L_r = s X_r \quad (7.16)$$

where X_r is the reactance of the leakage inductance at synchronous frequency or when the rotor is at standstill. Therefore, the equivalent circuit can be represented as in Figure 7.14.

Consider again the relationships between the various frequencies in the induction motor. The stator voltages and currents are operating at synchronous frequency. The rotor voltage and current, when viewed from the rotor, are operating at a frequency of slip times synchronous frequency. Since the rotor impedances are part of and rotating with the rotor, sf is the frequency that determines the leakage reactance of the rotor. However, the rotor is also turning at a speed that, when added to the rotor frequency, equals synchronous frequency. Therefore, all rotor voltages and currents appear to have a frequency of synchronous frequency when viewed from the stator.

This relationship means that for a certain operating speed, the interaction between the stator and rotor of an induction motor, when viewed from the stator, is very similar to the interaction of the primary and secondary of a transformer. This concept is illustrated in Figure 7.15. At any specific speed voltage E_1 and current I_2 are transformed to voltage E_r and current I_r on the rotor. Both of these rotor quantities appear to be at synchronous frequency from the stator frame of reference. However, if the speed of the motor changes, then the same E_1 and I_2 would be transformed to different values of voltage and current on the rotor. The rotor quantities would still appear to be operating at synchronous frequency when viewed from the stator. Therefore, the induction motor model, when viewed from the stator, is similar to a transformer model whose mutual inductance varies with speed. In addition, the rotor leakage inductance varies with speed. To make the model in Figure 7.15 useful, the ideal

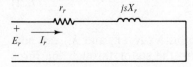

Figure 7.14 Rotor circuit with synchronous reactance.

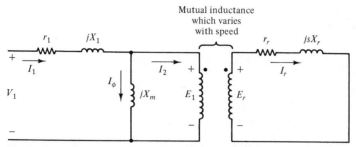

Figure 7.15 Equivalent circuit with ideal transformer.

transformer must be eliminated, and the rotor impedances will have to be represented on the stator side. For a given r_r and sX_r of the rotor, a voltage E_r will result in a certain current I_r. This phenomenon in turn will require a certain value of I_2 to flow in the stator. The problem here is to determine the impedance that, when connected directly across the stator-induced voltage E_1, will result in the same value of I_2 flowing. This impedance can then represent the rotor in the stator circuit.

The current in the rotor circuit is calculated by

$$I_r = \frac{E_r}{r_r + jsX_r} \tag{7.17}$$

In a previous chapter the MMFs on each side of an ideal transformer were shown to be balanced. Therefore, I_2 and I_r are related by Equation 7.18, when viewed from the stator so that both appear to be operating at synchronous frequency:

$$N_1 I_2 = N_2 I_r \tag{7.18}$$

Earlier voltage magnitudes were found to be related. Therefore, if voltages E_1 and E_r are viewed from the stator so that both appear to be operating at synchronous frequency, they have the relationship

$$\frac{E_1}{E_r} = \frac{N_1}{N_2 s} \tag{7.19}$$

Making the appropriate substitutions, Equation 7.17 becomes

$$\frac{N_1}{N_2} I_2 = \frac{(N_2/N_1)sE_1}{r_r + jsX_r} \tag{7.20}$$

Solving for I_2,

$$I_2 = \frac{sE_1}{a^2 r_r + jsa^2 X_r} \tag{7.21}$$

where a is the turns ratio of the stator and rotor, N_1/N_2. r_2 and X_2 can now be defined as the rotor impedances referred to the stator through the turns

7.4 THREE-PHASE INDUCTION MOTORS

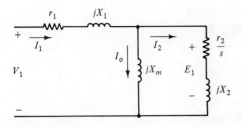

Figure 7.16 Three-phase induction motor equivalent circuit.

ratio. That is,

$$r_2 = a^2 r_r = \left(\frac{N_1}{N_2}\right)^2 r_r \qquad (7.22)$$

$$X_2 = a^2 X_r = \left(\frac{N_1}{N_2}\right)^2 X_r \qquad (7.23)$$

This substitution can now be made into Equation 7.21:

$$I_2 = \frac{sE_1}{r_2 + jsX_2} \qquad (7.24)$$

A more meaningful form of Equation 7.24 is available by dividing s into the numerator and denominator on the right-hand side:

$$I_2 = \frac{E_1}{(r_2/s) + jX_2} \qquad (7.25)$$

So, if the impedance, $(r_2/s) + jX_2$, where r_2 and X_2 are defined in Equations 7.22 and 7.23, is connected across voltage E_1, the same current I_2 will flow as when the rotor was explicitly taken into account. The equivalent circuit of one phase of a three-phase induction motor is given in Figure 7.16. The voltage V_1 is the line-to-neutral terminal voltage. Normally the impedances r_1, X_1, X_m, r_2, and X_2 are known values supplied by the manufacturer.

7.4.5 Analyzing the Equivalent Circuit

The equivalent circuit shown in Figure 7.16 can be used to determine the operating characteristics of an induction motor. The voltages, currents, power losses, and torques associated with various operating conditions can be calculated. For example, the power being consumed by a three-phase induction motor can be calculated as

$$P_{in} = 3V_1 I_1 (\text{pf}) \qquad (7.26)$$

where pf means power factor. The stator copper losses are

$$P_{cl1} = 3I_1^2 r_1 \qquad (7.27)$$

The power transferred across the air gap from the stator to the rotor is

$$P_g = P_{in} - P_{cl1} \qquad (7.28)$$

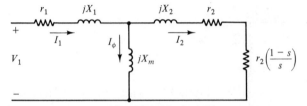

Figure 7.17 Equivalent circuit that explicitly represents the rotor circuit resistance.

The power transferred to the rotor can also be written

$$P_g = 3I_2^2 \frac{r_2}{s} \tag{7.29}$$

From Equation 7.22 the rotor resistance referred to the stator is equal to r_2. Therefore, the copper losses in the rotor are

$$P_{cl2} = 3I_2^2 r_2 \tag{7.30}$$

The resistance of the circuit representing the rotor in Figure 7.16 can be divided into two resistances such that one of them represents the rotor copper losses. This equivalent circuit is shown in Figure 7.17. The summation of the two rotor resistances still equals r_2/s:

$$r_2 + r_2 \frac{1-s}{s} = \frac{r_2 s}{s} + \frac{r_2 - r_2 s}{s} = \frac{r_2}{s} \tag{7.31}$$

The internal mechanical power developed or the power that tends to move the rotor is

$$P_d = P_g - P_{cl2}$$
$$= 3I_2^2 \frac{r_2}{s} - 3I_2^2 r_2$$
$$= 3I_2^2 r_2 \left(\frac{1-s}{s}\right)$$
$$= P_g(1-s) \tag{7.32}$$

Therefore, the other rotor resistance in the equivalent circuit of Figure 7.17 represents the internal mechanical power developed.

The output power of the motor can be determined by subtracting windage, friction, and core losses (designated P_l) from the internal mechanical power developed:

$$P_{out} = P_d - P_l \tag{7.33}$$

The internal electromagnetic torque developed, τ_d, and the output torque τ_{out} can be determined from the calculated values of power:

$$\tau_{out} = \frac{P_{out}}{\omega_m} = \frac{P_{out}}{(1-s)\omega_s} \tag{7.34}$$

7.4 THREE-PHASE INDUCTION MOTORS

$$\tau_d = \frac{P_d}{\omega_m} = \frac{3I_2^2 r_2[(1-s)/s]}{(1-s)\omega_s} = \frac{3I_2^2(r_2/s)}{\omega_s} = \frac{P_g}{\omega_s} \quad (7.35)$$

where ω_m is the mechanical angular velocity and ω_s is the synchronous angular velocity.

EXAMPLE 7.5

A 60-Hz, three-phase, four-pole, 220-V induction motor has the following impedances referred to the stator:

$r_1 = 0.2\,\Omega \quad X_1 = 0.5\,\Omega \quad r_2 = 0.15\,\Omega \quad X_2 = 0.3\,\Omega \quad X_m = 16.0\,\Omega$

If the motor is operating at a slip of 3 percent, calculate the speed of the motor in rpms, stator current, power across the air gap, power output, torque output, and efficiency. Friction, windage, and core losses total 470 W.

The equivalent circuit is

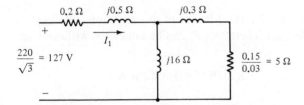

Synchronous speed in rpm equals

$$n_s = \left(\frac{2}{p}\right) f (60 \text{ sec/min}) = \tfrac{2}{4}(60)60 = 1800 \text{ rpm}$$

Rotor speed equals

$$n = n_s(1-s) = 1800(0.97) = 1746 \text{ rpm}$$

To find I_1, the circuit must be reduced by combining impedances. The parallel combination of jX_m and $(r_2/s) + jX_2$ will be defined as Z_f:

$$Z_f = \frac{(r_2/s + jX_2)jX_m}{r_2/s + j(X_2 + X_m)} = \frac{(5 + j0.3)(j16)}{5 + j16.3}$$
$$= 4.70\,\underline{/20.5°} = 4.40 + j1.65$$

If the terminal voltage is chosen as reference, the equivalent circuit now appears as

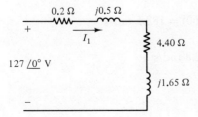

By combining Z_f and $r_1 + jX_1$, one can find I_1:

$$I_1 = \frac{127/0°}{4.60 + j2.15} = 25.0/-25.1° \text{ A}$$

The power across the air gap is defined as

$$P_g = 3I_2^2 \frac{r_2}{s}$$

Therefore, by knowing I_1 and using the current divider rule, one can find I_2 and P_g. However, if the reduced model with Z_f is to be equivalent to the original equivalent circuit with r_2/s, then the power consumed by Z_f must equal the power consumed by r_2/s. Therefore,

$$3I_2^2 \frac{r_2}{s} = 3I_1^2 r_f$$

where r_f is the real part of Z_f. The left side of this equation is defined as the power across the air gap. Therefore, P_g can be calculated without finding I_2:

$$P_g = 3I_1^2 r_f = 3(25.0)^2(4.40) = 8250 \text{ W}$$

Power output equals

$$P_{out} = P_d - P_l$$

and

$$P_d = P_g(1-s) = 8250(0.97) = 8003 \text{ W}$$

Therefore,

$$P_{out} = P_d - P_l = 8003 - 470 = 7533 \text{ W}$$

The torque output is

$$\tau_{out} = \frac{P_{out}}{\omega_m}$$

The mechanical angular speed equals

$$\omega_m = (1-s)\omega_s$$

where the synchronous angular speed equals

$$\omega_s = \left(\frac{2}{p}\right)2\pi f = \left(\frac{2}{4}\right)2\pi(60) = 188.5 \text{ rad/s}$$

so

$$\omega_m = (0.97)(188.5) = 182.8 \text{ rad/s}$$

and

$$\tau_{out} = \frac{7533}{182.8} = 41.2 \text{ N·m}$$

7.4 THREE-PHASE INDUCTION MOTORS

The efficiency is defined as power out divided by power in. The power input is calculated as

$$P_{in} = 3V_1 I_1 (\text{pf}) = 3(127)(25.0)\cos(-25.1°) = 8626 \text{ W}$$

Therefore,

$$\text{efficiency} = \frac{7533}{8626} = 0.873 \quad \text{or} \quad 87.3\%$$ ■■

7.4.6 The Equivalent Circuit at Start-up

At start-up the rotor is at standstill and the slip of the motor is 1.0. If all the values of slip in the model of Figure 7.10 are set to a value of 1, the equivalent circuit that results will be for the induction motor at start-up. All powers and torques can be calculated as shown previously, except the output quantities. Since the motor is at standstill, no windage or friction losses exist. In addition, ω_m will be 0, so τ_{out} will be undefined. However, τ_d can be calculated for $s = 1.0$ and is called the starting torque.

EXAMPLE 7.6

For the motor in Example 7.5, calculate the starting torque and the stator current at start-up.

The equivalent circuit is

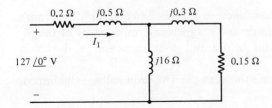

Combining the rotor and magnetizing circuits,

$$Z_f = \frac{(0.15 + j0.3)(j16)}{0.15 + j16.3} = 0.329 \underline{/63.9°} = 0.145 + j0.295 \text{ }\Omega$$

Z_f is now added to the stator impedances:

$$Z_f + (r_1 + jx_1) = 0.345 + j0.795 \text{ }\Omega$$

The stator current can now be calculated:

$$I_1 = \frac{127\underline{/0°}}{0.345 + j0.795} = 147\underline{/-66.5°} \text{ A}$$

Notice that the starting current is almost six times the load current calculated in Example 7.5. The power across the air gap equals

$$P_g = 3I_1^2 R_f = 3(147)^2(0.145) = 9400 \text{ W}$$

The starting torque equals

$$\tau_d = \frac{P_g}{\omega_s} = \frac{9400}{188.5} = 49.9 \text{ N}\cdot\text{m}$$

■ ■

7.4.7 Determining Power and Torque from the Thevenin Equivalent

All the impedances of the three-phase induction motor equivalent circuits are constant with the exception of those containing slip terms. The r_2/s term in Figure 7.16 and the $r_2[(1-s)/s]$ term in Figure 7.17 vary with the speed of the motor. If these equivalent circuits could be reduced in a manner so as to retain the identity of these impedances, then analyses of several different operating conditions could be done with a single reduction. The method of analysis used in Example 7.5 would require a circuit reduction for each slip to be analyzed.

One method that would retain the rotor resistance is the application of Thevenin's theorem. Thevenin's theorem allows a network of linear impedances and voltage sources to be represented by a single voltage source and a single impedance as viewed from two terminals. The equivalent voltage source is the voltage that appears across these terminals when the terminals are open-circuited. The equivalent impedance equals the impedance seen from the terminals looking into the network to be reduced with all voltage sources short-circuited.

In order to retain the rotor resistances in Figures 7.16 and 7.17, Thevenin's theorem will be applied to the stator and magnetizing impedances of the induction motor models. The circuit before and after reduction is shown in Figure 7.18.

From application of Thevenin's theorem, the Thevenin voltage and impedance are

$$V_{\text{Th}} = V_1 \frac{jX_m}{r_1 + j(X_1 + X_m)} \tag{7.36}$$

$$Z_{\text{Th}} = r_{\text{Th}} + jX_{\text{Th}} = \frac{(r_1 + jX_1)(jX_m)}{r_1 + j(X_1 + X_m)} \tag{7.37}$$

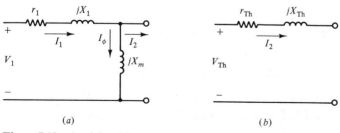

Figure 7.18 Applying Thevenin's theorem.

7.4 THREE-PHASE INDUCTION MOTORS

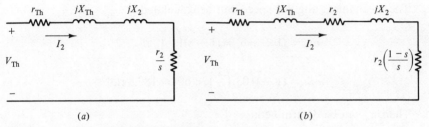

Figure 7.19 Thevenin equivalent circuit.

The equivalent circuits of the induction motor shown in Figures 7.16 and 7.17 now appear as illustrated in Figure 7.19.

EXAMPLE 7.7

Find the internal mechanical power developed and internal electromagnetic torque developed for the motor of Example 7.5 using the Thevenin equivalent model at slips of 3 and 5 percent.

The Thevenin voltage and impedance are

$$V_{Th} = 127\underline{/0°} \frac{j16}{0.2 + j16.5} = 123.1\underline{/0.7°} \text{ V}$$

$$Z_{Th} = \frac{(0.2 + j0.5)(j16)}{0.2 + j16.5} = 0.188 + j0.487 \text{ }\Omega$$

The equivalent circuit can now be drawn:

At a slip of 3 percent the total input impedance is

$$Z_{in} = \left(0.188 + 0.15 + 0.15\frac{1 - 0.03}{0.03}\right) + j(0.487 + 0.3) = 5.26\underline{/8.6°} \text{ }\Omega$$

The current I_2 is

$$I_2 = \frac{123.1\underline{/0.7°}}{5.25\underline{/8.6°}} = 23.44\underline{/-7.9} \text{ A}$$

Therefore,

$$P_d = 3I_2^2 r_2 \left(\frac{1-s}{s}\right) = 3(23.44)^2(0.15)\left(\frac{1 - 0.03}{0.03}\right) = 7999 \text{ W}$$

The mechanical angular speed can be calculated:

$$\omega_m = (1-s)\omega_s = (1-s)\left(\frac{2}{p}\right)2\pi f$$

$$= (1-0.03)\left(\frac{2}{4}\right)2\pi(60) = 182.8 \text{ rad/s}$$

Then, τ_d can be determined:

$$\tau_d = \frac{P_d}{\omega_m} = 43.8 \text{ N·m}$$

Doing the same calculations for a slip of 5 percent, we find

$$Z_{in} = \left(0.188 + 0.15 + 0.15\frac{1-0.05}{0.05}\right) + j(0.487 + 0.3)$$

$$= 3.28\underline{/13.9°} \ \Omega$$

$$I_2 = \frac{123.1\underline{/0.7°}}{3.28\underline{/13.9°}} = 37.53\underline{/-13.2°} \text{ A}$$

$$P_d = 3(37.53)^2(0.15)\left(\frac{1-0.05}{0.05}\right) = 12{,}042 \text{ W}$$

$$\omega_m = (1-0.05)\left(\frac{2}{4}\right)2\pi(60) = 179.1 \text{ rad/s}$$

$$\tau_d = \frac{12{,}042}{179.1} = 67.2 \text{ N·m}$$

∎ ∎

7.4.8 Calculation of Maximum Torque

The maximum or breakdown internal electromagnetic torque can be easily calculated using the Thevenin equivalent. From Equation 7.35 the internal torque developed equals

$$\tau_d = \frac{P_g}{\omega_s} \quad (7.38)$$

The torque will be a maximum when P_g is a maximum. From the Thevenin equivalent circuit the magnitude of I_2 can be found

$$I_2 = \frac{V_{Th}}{[(r_{Th} + r_2/s)^2 + (X_{Th} + X_2)^2]^{1/2}} \quad (7.39)$$

Therefore,

$$P_g = 3I_2^2\frac{r_2}{s} = 3\frac{V_{Th}^2}{(r_{Th} + r_2/s)^2 + (X_{Th} + X_2)^2}\frac{r_2}{s} \quad (7.40)$$

To find at what value of the variable r_2/s the maximum P_g occurs, the deriva-

7.4 THREE-PHASE INDUCTION MOTORS

tive of the right-hand side of Equation 7.40 with respect to r_s/s must be calculated and set equal to 0.

$$0 = \frac{3V_{Th}^2[r_{Th}^2 - (r_2/s)^2 + (X_{Th} + X_2)^2]}{[(r_{Th} + r_2/s)^2 + (X_{Th} + X_2)^2]^2} \quad (7.41)$$

The numerator of Equation 7.41 must be equal to 0. Therefore,

$$0 = r_{Th}^2 - \left(\frac{r_2}{s}\right)^2 + (X_{Th} + X_2)^2 \quad (7.42)$$

Equation 7.42 is a form of the familiar impedance-matching technique in circuit theory. Maximum power transfer to r_2/s will occur when its magnitude equals the magnitude of the other impedances in the circuit:

$$\frac{r_2}{s} = \sqrt{r_{Th}^2 + (X_{Th} + X_2)^2} \quad (7.43)$$

Therefore, P_g and τ_d will be a maximum when

$$s_{\max\tau} = \frac{r_2}{\sqrt{r_{Th}^2 + (X_{Th} + X_2)^2}} \quad (7.44)$$

If this value of slip is substituted into Equations 7.40 and 7.38, Equation 7.45 results:

$$\tau_{d\max} = \frac{3}{\omega_s} \frac{V_{Th}^2 \sqrt{r_{Th}^2 + (X_{Th} + X_2)^2}}{[r_{Th} + \sqrt{r_{Th}^2 + (X_{Th} + X_2)^2}]^2 + (X_{Th} + X_2)^2} \quad (7.45)$$

Equation 7.45 reduces to

$$\tau_{d\max} = \frac{3}{2\omega_s} \frac{V_{Th}^2}{r_{Th} + \sqrt{r_{Th}^2 + (X_{Th} + X_2)^2}} \quad (7.46)$$

EXAMPLE 7.8

Find the maximum torque and the slip at which maximum torque occurs for the motor in Example 7.5.

From Example 7.7 the values of V_{Th} and Z_{Th} are

$$V_{Th} = 123.1\underline{/0.7°}\text{ V}$$
$$Z_{Th} = r_{Th} + jX_{Th} = 0.188 + j0.487 \text{ }\Omega$$

The maximum torque can be calculated from Equation 7.46:

$$\tau_{d\max} = \frac{3}{2(188.5)} \frac{(123.1)^2}{0.188 + \sqrt{(0.188)^2 + (0.487 + 0.3)^2}} = 120.9 \text{ N·m}$$

The slip at which maximum torque occurs can be calculated from Equation 7.44:

$$s_{\max\tau} = \frac{0.15}{\sqrt{(1.88)^2 + (0.487 + 0.3)^2}} = 0.185 \quad \text{or} \quad 18.5 \text{ percent} \quad \blacksquare\blacksquare$$

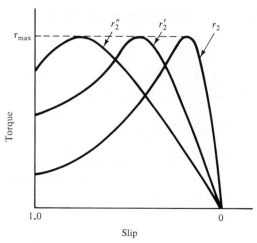

Figure 7.20 Variation of slip at maximum torque with r_2.

Note that the maximum torque is not a function of r_2. However, the slip at which the maximum torque occurs varies directly with r_2. Figure 7.20 illustrates this concept where $r_2 < r_2' < r_2''$.

7.4.9 Varying Rotor Resistance and Starting Torque

As discussed previously, motors are usually designed so that a large variance in torque can be achieved within a small range of speed. Figure 7.20 shows that a low value of rotor resistance would be desirable. However, the starting torque increases as rotor resistance increases. A large starting torque is usually desirable, so a higher value of rotor resistance would be needed. This paradox can be overcome by having a large rotor resistance at start-up and a lower resistance at normal operating conditions.

In a wound rotor machine direct access to the rotor circuits is available through the slip rings. An external resistance can be connected in series with each rotor coil while the motor is being started and removed as operating speed is approached.

A squirrel cage rotor is not directly accessible. The design of the motor must incorporate the ability of the rotor resistance to vary with speed. This goal is quite often accomplished by making the rotor bars narrow and deep, as shown in Figure 7.21. If the rotor bar is thought of as being divided into many small sections, the current in each section will create a leakage flux as shown in Figure 7.21b. The flux lines must cross the rotor bar at the top of their paths. However, on the bottom of all the flux paths, the flux will travel through the low reluctance iron below the rotor bar. Therefore, the segments at the bottom of the bar will have more flux linkages than the segments at the top. The leakage inductance of the bottom segments will be significantly higher than the top segments.

7.5 SINGLE-PHASE INDUCTION MOTORS

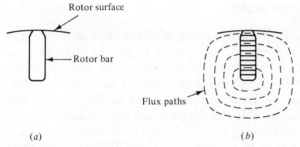

Figure 7.21 Deep rotor bars.

At start-up the rotor frequency is 60 Hz. The inductive leakage reactance of each segment will be a 60-Hz value and will be the major part of the rotor impedance. Since the reactance at the bottom of the bar will be much higher than the reactance at the top, the current will tend to flow in the top segments of the rotor bar. The current will be flowing in a small cross section of the conductor, and the effective rotor resistance will be a relatively high value.

As the rotor speeds up, the rotor frequency decreases. At normal operating speed the rotor frequency is usually in the range of 2 to 5 Hz. At these frequencies the rotor reactance will be much less, and the resistance will dominate the rotor impedance. Therefore, the current will distribute more evenly throughout the rotor bar. A more even current distribution means a larger cross-sectional area of the conductor will be available for the current to flow and the effective resistance of the rotor will be small.

These conditions match exactly with the desired design goals. A high rotor resistance at start-up will result in a high starting torque, and a low resistance at operating speeds will give good full-load operating characteristics.

7.5 SINGLE-PHASE INDUCTION MOTORS

One of the most common types of customer loads is single-phase induction motors. These motors are used extensively in residential and commercial load centers. They are usually designed and constructed for fractional horsepower duty, and a wide variety of motor designs exist to suit special operating requirements. Their most common uses are for fans (e.g., home furnance fans) and compressors (e.g., refrigerator compressor).

Earlier in this chapter power system loads were identified as three-phase devices. However, three-phase electric service is rarely provided for individual residential loads and is also not common for commercial loads. Most of these load centers receive only single-phase electric service, which is supplied through step-down transformers. The load side on these transformers is usually center tapped to provide two sets of voltages for the load. Thus, the motors used at single-phase loads must be single-phase devices that start with only single-phase electrical excitation and operate as smoothly and quietly as possible.

7.5.1 Magnetic Field of the Stator

Earlier in this chapter the need for a rotating magnetic field in the air gap of motors was identified. The rotor MMF is made to try and align with the rotating field, and in doing so a torque develops that causes rotation of the rotor. In three-phase induction and synchronous motors, the rotating MMF on the stator is established by using distributed turns in the windings and balanced three-phase excitation. However, single-phase motors have only one set of windings and only one phase of current excitation. Therefore, creating a rotating MMF must be achieved by other means. Figure 7.22 shows the basic construction of a single-phase induction motor. The winding on the stator is assumed to be symmetrically distributed about its magnetic axis, indicated by the vertical line through the center of the machine. Current i_1 flows in on the right side and out on the left side.

Chapter 3 derived the MMF directed radially outward for each phase of a three-phase stator. The results of that derivation can be applied directly by recognizing that Figure 7.22 shows the winding distribution for phase a of the three-phase stator. Thus, the radially directed outward MMF for the single-phase stator is

$$F_1(\psi) = N_1 i_1 \cos(\psi) \tag{7.47}$$

If sinusoidal current is used in the windings such that

$$i_1(t) = I_{1m} \cos \omega t \tag{7.48}$$

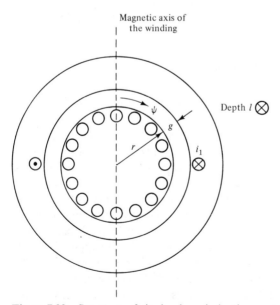

Figure 7.22 Structure of single-phase induction motor.

7.5 SINGLE-PHASE INDUCTION MOTORS

$F_1(\psi)$ can be written

$$F_1(\psi) = N_1 I_{1m} \cos \omega t \cos(\psi) \tag{7.49}$$

Equation 7.49 shows that $F_1(\psi)$ for single-phase motors does not rotate in the same sense as the three-phase machine. That is, its peak magnitude point is pointed either up or down when referenced to the magnetic axis of Figure 7.22 and its magnitude

$$|F_1(\psi)| = N_1 I_{1m} \cos \omega t \tag{7.50}$$

varies with time. This kind of MMF is sometimes referred to as a "breathing field" because it expands and contracts in the same place on the stator. Thus, a dilemma appears to exist in the operation of a single-phase machine. Its winding does not provide a rotating magnetic field for the rotor MMF to chase.

The dilemma of the breathing field on the stator can be overcome by resolving $F_1(\psi)$ into two components using a trigonometric identity:

$$\begin{aligned} F_1(\psi) &= N_1 I_{1m} \left[\frac{\cos(\omega t - \psi)}{2} + \frac{\cos(\omega t + \psi)}{2} \right] \\ &= \tfrac{1}{2} N_1 I_{1m} \cos(\omega t - \psi) + \tfrac{1}{2} N_1 I_{1m} \cos(\omega t + \psi) \end{aligned} \tag{7.51}$$

Each term in Equation 7.51 can be thought of as a rotating MMF with a magnitude of half of $F_1(\psi)$. The first term is identical in form with $F_s(\psi)$ developed in Chapter 3. It is rotating forward with respect to the forward direction of ψ. Correspondingly, the second term is a backward-rotating field.

The forward- and backward-rotating MMFs both produce induction motor action. That is, they both produce a torque on the rotor. However, the torques caused by the fields act on the rotor in opposing directions such that their algebraic sum is 0 when the rotor is at standstill. Figure 7.23 illustrates this

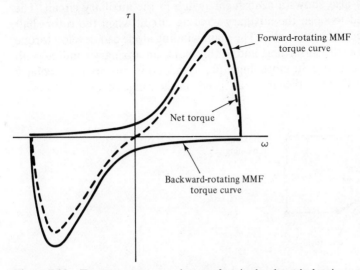

Figure 7.23 Torque versus speed curve for single-phase induction motor.

situation by plotting and summing the torque versus speed curves for both the forward-rotating and backward-rotating MMFs.

7.5.2 Capacitive Starting Single-Phase Induction Motors

Note that at rotor speeds other than zero, the motor is able to develop a torque. If the rotor speed is greater than zero, the torque is in a direction that will sustain forward rotation. If the rotor speed is less than zero, the torque is in a direction that will sustain backward rotation. This characteristic is one of the more interesting features of single-phase motors. If the rotor shaft can somehow be made to rotate either forward or backward, the motor will develop torque that sustains rotation. However, the single-phase induction motor is not able to start rotation from standstill.

This feature poses a design problem that must be overcome in order to make single-phase induction motors of any use. One solution to this problem is to add an auxiliary winding to the stator that is spatially oriented in quadrature with the original main winding. The auxiliary winding is connected to an electric power source in parallel with the main winding. Its impedance is designed such that its current is significantly out of phase with the main winding current. Figure 7.24 shows a schematic diagram of these windings.

The diagram shows that the impedance of the auxiliary winding is determined by L_a and C. The presence of the capacitor in the auxiliary winding will cause I_a to be phase shifted by an angle approaching 90° from I_m. Since both the auxiliary winding turns and current are approximately 90° out of phase with the main winding turns and current, the motor shown in Figure 7.24 behaves as if it were a two-phase induction motor. Two-phase excitation will develop a rotating air gap magnetic field. This field will cause a constant torque that will start the rotor moving from standstill.

Figure 7.24 also shows a centrifugal switch in the auxiliary circuit. The switch is designed to open the auxiliary winding circuit when the rotor shaft speed is some predetermined value. The main winding alone can develop torque once the rotor has started turning, and at that point an auxiliary winding is no longer needed. Therefore, in most single-phase induction motors the auxiliary winding is switched out of the circuit at speeds in the range of 60 to 80 percent of rated speed.

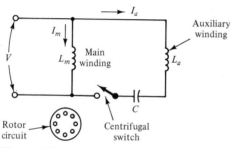

Figure 7.24 Schematic diagram of single-phase induction motor.

7.5.3 Split-Phase Single-Phase Induction Motors

Several different variations on the single-phase induction motor shown in Figure 7.24 are also in use. One of these variations is called a split-phase motor and is different from the capacitor start motor in that the capacitor is left out of the auxiliary winding. The split-phase motor uses an auxiliary winding that has an X/r ratio significantly less than the main winding X/r ratio. The result of this design feature is that the auxiliary winding current is out of phase with respect to the main winding current by a significant amount. As in the capacitor start motor, this phase difference makes the motor approximate the performance of a balanced two-phase motor.

7.5.4 Equivalent Circuit of a Rotating Single-Phase Motor

An equivalent electric circuit can be created for a single-phase induction motor by developing equivalent circuits for the forward- and backward-rotating MMFs and combining them. Recall that the development of an equivalent circuit for a three-phase induction motor was centered around the stator voltage E_1 induced by the resultant air gap flux ϕ. The single-phase induction motor has a similar voltage induced on its stator. This single-phase induction motor-induced voltage can be resolved into voltages induced by the forward-rotating flux wave and the backward-rotating flux wave. Therefore, E_1 for a single-phase motor has two components, E_{1F} and E_{1B}, that represent voltages induced by the forward-rotating flux wave and backward-rotating flux wave, respectively. Figure 7.25 shows the equivalent circuit for the stator using E_{1F} and E_{1B}. The quantities r_1 and X_1 are the stator resistance and leakage reactance of the stator, respectively.

Recall that in the three-phase motor, current I_1 was split into two components, I_ϕ and I_2. Component I_ϕ was the magnetizing component, and I_2 was the load component. Current components analogous to these components can be developed for both the forward- and backward-rotating circuit of the single-phase motor equivalent circuit. Figure 7.26 shows the first part of this development. Note carefully that the magnetizing reactance X_m has been divided by 2. At standstill the forward- and backward-induced voltages in a single-phase motor are of equal magnitude. The total induced voltage is the sum of E_{1F} and E_{1B}. If X_m is the total magnetizing reactance, then to obtain the actual

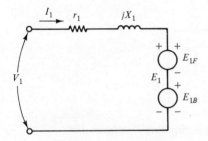

Figure 7.25 Equivalent stator circuit for a single-phase induction motor.

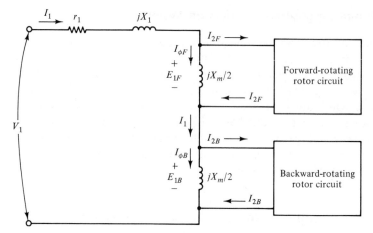

Figure 7.26 Identification of I_ϕ and I_2 for the single-phase motor.

magnetizing current, $X_m/2$ must be placed across each of the forward- and backward-induced voltages.

Figure 7.26 also shows two load component currents I_{2F} and I_{2B} flowing into the forward-rotating and backward-rotating equivalent rotor circuits, respectively. The equivalent rotor circuits are analogous to the rotor circuit developed and shown in Figures 7.14 to 7.16. They consist of resistance and leakage reactance and are short-circuited. They appear in the single-phase motor equivalent circuit, as shown in Figure 7.27. Notice that as in Figure 7.26 the reduced magnitude of forward- and backward-rotating MMFs and induced voltages is compensated by the factor of $\frac{1}{2}$ in the rotor circuit impedances. Also note that r_2 in the backward-rotating rotor equivalent is divided by $2 - s$. The

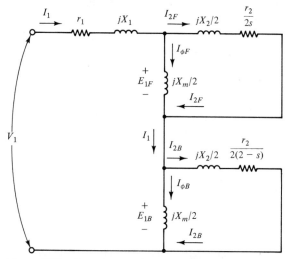

Figure 7.27 Complete equivalent circuit for a rotating single-phase motor.

7.5 SINGLE-PHASE INDUCTION MOTORS

reason for dividing by $2 - s$ is easily derived by recalling the definition of slip:

$$s_F = \frac{\omega_s - \omega_m}{\omega_s} \qquad (7.52)$$

where s_F = slip of a forward-rotating motor
ω_s = synchronous frequency of rotation of stator magnetic field
ω_m = mechanical frequency of rotor rotation

For a backward-rotating field the slip is defined as

$$s_B = \frac{-\omega_s - \omega_m}{-\omega_s} \qquad (7.53)$$

Solving for ω_m in Equation 7.52 and substituting in Equation 7.53, yields

$$s_B = \frac{-\omega_s - \omega_s + \omega_s s_F}{-\omega_s}$$

$$= 2 - s_F \qquad (7.54)$$

The calculation of currents, voltages, and powers are illustrated in the following examples.

EXAMPLE 7.9

Given a single-phase induction motor with the following characteristics, calculate the value of current I_1:

$r_1 = 2.5\ \Omega \qquad X_1 = 3.0\ \Omega \qquad r_2 = 2.5\ \Omega \qquad X_2 = 3.0\ \Omega$
$X_m = 60\ \Omega \qquad V_1 = 115\ V \qquad s = 0.05$

The equivalent circuit is shown in Figure 7.28. The calculation of I_1 is performed using Ohm's law:

$$I_1 = \frac{V_1}{Z_{motor}}$$

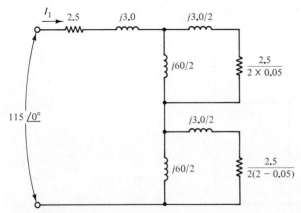

Figure 7.28 Equivalent circuit.

where Z_{motor} is the equivalent impedance of the circuit in Figure 7.28:

$$Z_{motor} = 23.97\underline{/44.86°}$$

Now calculate I_1:

$$I_1 = 115\underline{/0°}/23.97\underline{/44.86°}$$
$$= 4.797\underline{/-44.86°} \text{ A}$$

or
$$I_1 = 3.400 - j3.384 \text{ A}$$ ■■

EXAMPLE 7.10

For the motor given in Example 7.9, calculate the voltage induced on the main winding.

The voltages induced on the main windings of the forward- and backward-rotating motors are shown in Figure 7.29 as E_{1F} and E_{1B}, respectively. From the principle of superposition, the total voltage induced on the main winding is

$$E_1 = E_{1F} + E_{1B}$$
$$= I_m \frac{Z_F}{2} + I_m \frac{Z_B}{2}$$

where
$$\frac{Z_F}{2} = \frac{(jX_m/2)(r_2/2s + jX_2/2)}{r_2/2s + j(X_m/2 + X_2/2)}$$

and
$$\frac{Z_B}{2} = \frac{(jX_m/2)(r_2/2(2-s) + jX_2/2)}{r_2/2(2-s) + j(X_m/2 + X_2/2)}$$

From Example 7.9

$$I_1 = 3.4 - j3.384 \text{ A}$$
$$Z_F/2 = 13.91 + j12.47 \text{ }\Omega$$
$$Z_B/2 = 0.580 + j1.440 \text{ }\Omega$$

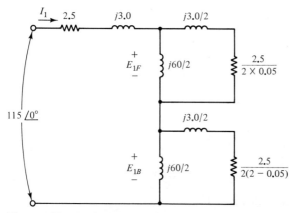

Figure 7.29 Equivalent circuit showing E_{MF} and E_{MB}.

7.5 SINGLE-PHASE INDUCTION MOTORS

The calculation of E_1 follows as

$$E_1 = (3.4 - j3.384)(13.91 + j12.47) + (3.4 - j3.384)(0.580 + j1.440)$$
$$= [89.49 + j(-4.666)] + (6.849 + j2.930)$$
$$= 96.34 - j1.736$$

or $\quad E_1 = 96.35 / -1.032°$ V ■■

EXAMPLE 7.11

For the motor given in Example 7.9, calculate its total air gap power, rotor copper loss, and developed mechanical power.

Recall that the air gap power per phase for a three-phase machine is calculated as

$$P_g = I_1^2 r_f$$

where I_1 = stator current
$r_f = \text{Re}\{Z_F\}$

In the single-phase motor, two components of air gap power exist. They are

$$P_{gF} = I_1^2 \frac{r_F}{2} \qquad P_{gB} = I_1^2 \frac{r_B}{2}$$

where P_{gF} = air gap power for the forward-rotating motor
P_{gB} = air gap power for the backward-rotating motor

$$\frac{r_F}{2} = \text{Re}\left\{\frac{Z_F}{2}\right\} \qquad \frac{r_B}{2} = \text{Re}\left\{\frac{Z_B}{2}\right\}$$

The total air gap power for the single-phase motor is

$$P_g = P_{gF} - P_{gB}$$

Here P_{gB} has been subtracted from P_{gF} because its backward rotation represents a drag or additional load on the forward-rotating shaft:

$$P_g = I_1^2 \frac{r_F}{2} - I_1^2 \frac{r_B}{2}$$
$$= (4.797)^2(13.91) - (4.797)^2(0.580)$$
$$= 306.7 \text{ W}$$

Recall that the rotor copper losses per phase are defined for a three-phase machine as

$$P_{cl2} = sP_g$$

For the single-phase machine this calculation is

$$P_{cl2} = s(P_{gF} - P_{gB})$$
$$= 0.05(306.7)$$
$$= 15.34 \text{ W}$$

The developed mechanical power is defined for a three-phase machine as

$$P_d = (1 - s)P_g$$

For the single-phase machine the developed mechanical power is

$$\begin{aligned}P_d &= (1 - s)(P_{gF} - P_{gB}) \\ &= (1 - 0.05)(306.7) \\ &= 291.4 \text{ W}\end{aligned}$$

∎∎

7.6 DIRECT-CURRENT MOTORS

Direct-current motors have a unique place in an alternating current world. They have superior torque and speed range capabilities as compared to synchronous and induction motors. For this reason they are selected for use in applications requiring these characteristics, such as rolling mills, power shovels, and railroad locomotives.

Figure 7.30 Rotor and stator of a dc motor.

7.6 DIRECT-CURRENT MOTORS

7.6.1 Construction of dc Motors

As was discussed earlier in this chapter, the basic design of a dc motor tries to keep the magnetic field and rotating coil MMF stationary at 90° apart. In order to keep the coil MMF stationary, the rotor of a dc motor has several turns of wire or coils distributed in slots around its periphery. As the rotor rotates, the direction of current flow in the coils is switched so that the coil MMF remains oriented in a constant direction. The device that does the switching is the commutator and is attached to the rotor. Figure 7.30 shows a dc motor rotor removed and sitting next to its stator. Note the commutator at the end of the rotor. It consists of a collection of copper segments placed parallel to the rotor shaft. Each segment is insulated from its neighbors and is attached to specific coil terminals.

Figure 7.30 also shows the components called brushes. During normal operation, each brush is in contact with one or more commutator segments as the rotor spins. The brush provides the electrical path of current flow from the motor power supply to the rotor coils. Figure 7.31 shows the same dc motor shown in Figure 7.30 but with the rotor in its normal operating position inside the stator. The brushes are shown pressing against the commutator. They are spring loaded such that they maintain firm contact with the commutator segments. Figure 7.32 shows the same motor but with the view looking straight into the rotor shaft and at the brush rigging.

Figure 7.31 Rotor commutator and brush contact inside a dc motor.

Figure 7.32 Rotor commutator and brush contact view looking into the rotor shaft.

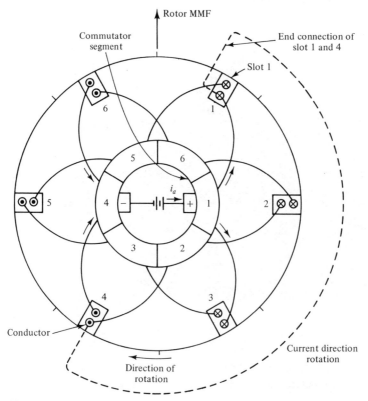

Figure 7.33 Brushes in contact with commutator segments 1 and 4.

7.6 DIRECT-CURRENT MOTORS

The function of the brush commutator assembly is to switch the direction of current flow in the rotor windings such that the rotor MMF remains stationary with respect to the stator winding field. Figures 7.33 and 7.34 show the rotor winding–commutator connections and illustrate how the commutator switches the rotor current flow direction. In Figure 7.33 the brushes are in contact with commutator segments 1 and 4. This rotor position results in current flow into the page in the slot conductors on the right side of the rotor and current flow out of the page in the slot conductors on the left side of the rotor. Thus, the rotor MMF has the orientation as shown.

Figure 7.34 shows the same rotor a short time later, during which it has rotated to a position such that the brushes are in contact with commutator segments 3 and 6. The direction of current flow for slot conductors on the right and left sides of the rotor is the same as it was in Figure 7.33, and as a result, the rotor MMF maintains its previous orientation. However, conductors in slot 6 and slot 4 have experienced a reversal of current direction during the change in rotor position from Figure 7.33 to 7.34. This current reversal has occurred

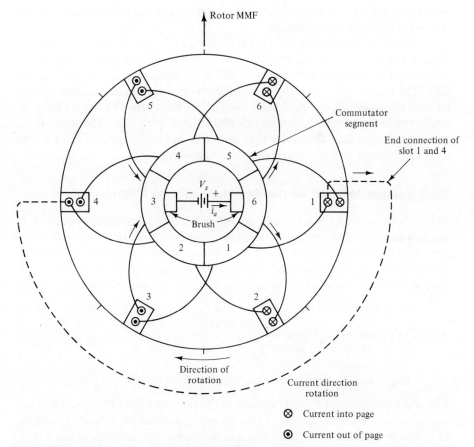

Figure 7.34 Brushes in contact with commutator segments 3 and 6.

Figure 7.35 View of the stator field coils of a dc motor.

because of the change in position of the commutator segments relative to the position of the stationary brushes.

The rotor MMF shown in Figures 7.33 and 7.34 has a vertical orientation. As was noted earlier in this chapter, the most desirable position of the stator field would be 90° apart from the rotor MMF or horizontal in Figures 7.33 and 7.34. The stator field is usually established with a set of coils wound around an iron assembly called a field pole. Figure 7.35 shows the field coils and poles inside the stator of the dc motor introduced in Figure 7.30. The view of the stator is looking into the cavity that normally holds the rotor. The two field coils can be seen at the top and bottom of the cavity.

7.6.2 Voltage Induced on the Armature of a dc Motor

The coils located on the rotor of a dc motor are more traditionally referred to as the armature windings; the entire rotor assembly is referred to as the armature. During normal motor operation, the armature windings are spinning within a magnetic field and thus have a voltage induced on them in accordance with Faraday's law. This voltage can be expressed in the following manner:

$$E_g = K_a \phi \omega_m \quad (7.55)$$

where E_g = armature voltage, V
  K_a = armature constant
  ϕ = magnetic flux per pole, Wb
  ω_m = speed of rotor rotation, rad/s

The armature constant K_a is a function of the rotor coil assembly geometry and can be calculated if the details of this geometry are known. However, calculations involving Equation 7.55 are more easily performed using the product $K_a \phi$, which can be calculated from the magnetization curve for a dc machine.

7.6 DIRECT-CURRENT MOTORS

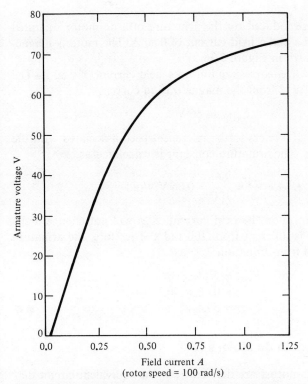

Figure 7.36 Magnetization curve for a dc machine.

The magnetization curve shows the relationship between speed of rotation, armature voltage, and field coil current. It is created from test data of a machine taken while driving the rotor at a constant speed with the armature open-circuited. The field current is varied over a range of values while recording the armature voltage. A plot of the armature voltage versus the field current yields the magnetization curve. Figure 7.36 shows a magnetization curve for a dc machine rotated at a speed of 100 rad/s.

EXAMPLE 7.12

Calculate the value of $K_a\phi$ for a dc motor operated at a speed of 100 rad/s and field current of 0.6 A. The motor's magnetization curve is shown in Figure 7.36.

The value of E_g for the given field current can be read from the magnetization curve as

$$E_g = 61 \text{ V}$$

Using Equation 7.55, the value of $K_a\phi$ is calculated as

$$K_a\phi = \frac{E_g}{\omega_m} = \frac{61 \text{ V}}{100 \text{ rad/s}} = 0.61 \text{ V·s/rad}$$

EXAMPLE 7.13

Calculate the voltage induced on the armature of a dc motor operated at a speed of 90 rad/s and a field current of 0.80 A. The motor's magnetization curve is shown in Figure 7.36.

First solve for $K_a\phi$ corresponding to a field current of 0.80 A. The armature voltage is read from the magnetization curve:

$$E_{g100} = 66 \text{ V}$$

where the subscript 100 refers to the reference speed associated with the magnetization curve. The armature constant is calculated as

$$K_a\phi = \frac{E_g}{\omega} = \frac{66}{100} = 0.66 \text{ V·s/rad}$$

Since ϕ is dependent upon the field current, $K_a\phi$ will not change if the speed of the motor is different from 100 rad/s. Therefore, the armature voltage is calculated from Equation 7.55 as

$$\begin{aligned} E_{g90} &= K_a\phi\omega_m \\ &= 0.66 \times 90 \\ &= 59.4 \text{ V} \end{aligned}$$
∎∎

7.6.3 Power and Torque in dc Motors

Power calculations for dc motors are derived from the equivalent circuit diagram shown in Figure 7.37. The circuit shows an applied dc voltage V_T, which causes the armature current I_a to flow into the terminals of the motor. Resistance within the armature windings is shown as r_a, and the voltage induced on the armature is shown as E_g. The current supplied to the field windings is I_f and is shown flowing in a separate inductor.

The armature input power of the motor is calculated as

$$P_{\text{in}} = V_T I_a \tag{7.56}$$

Since the armature resistance dissipates some of this power, the power left for mechanical use appears as

$$P_d = P_{\text{in}} - I_a^2 r_a = E_g I_a \tag{7.57}$$

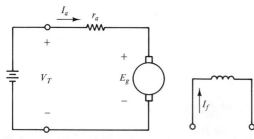

Figure 7.37 Equivalent circuit for a dc motor.

7.6 DIRECT-CURRENT MOTORS

where P_d is called the developed power. From the derived developed power expression, the development torque is derived as

$$\tau_d = \frac{P_d}{\omega_m} = \frac{E_g I_a}{\omega_m} \quad (7.58)$$

where ω_m is the angular velocity of the motor shaft in radians per second.

EXAMPLE 7.14

Calculate I_a, P_{in}, P_d, and τ_d for the motor in the previous example problem given the following motor characteristics:

$$V_T = 65 \text{ V} \quad r_a = 0.2 \text{ }\Omega$$

The equivalent circuit is

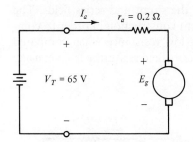

E_g was calculated as 59.4 V. From Kirchhoff's voltage law, the value of I_a is calculated as

$$I_a = \frac{V_T - E_g}{r_a}$$
$$= \frac{65 - 59.4}{0.2}$$
$$= 28 \text{ A}$$

The power and torque calculations follow:

$$P_{in} = V_T I_a$$
$$= (65)(28)$$
$$= 1820 \text{ W}$$

$$P_d = P_{in} - I_a^2 r_a$$
$$= 1820 - (28)^2(0.2)$$
$$= 1663.2 \text{ W}$$

$$\tau_d = \frac{P_d}{\omega_m}$$
$$= \frac{1663.2}{90}$$
$$= 18.48 \text{ N·m}$$

7.6.4 Field and Armature-Winding Connections

Since the field and armature windings of a dc motor are separate electric circuits, they can be connected in various ways that yield different performances for different operating requirements. Figure 7.38 shows four of these connections.

The separately excited connection shows the field winding powered from a separate dc source. In contrast, the shunt connection has the field winding connected in parallel with the armature. The current in the field circuit for the shunt connection is usually varied by a rheostat in series with it. The series-connected motor has its field winding placed in series with its armature winding. Finally, the cumulative compound motor combines the connections of the shunt and series motors. The shunt and series field windings are connected such that their magnetic fields are aligned in the same direction. This condition is illustrated in Figure 7.38 by the polarity dots on the shunt and series fields. The shunt and series windings are usually of different construction in order to coincide with their uses. The series winding has a low resistance when compared to the shunt winding. The design prevents the series field resistance from causing a large power dissipation within the motor.

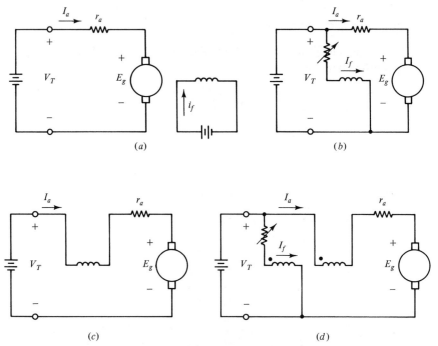

Figure 7.38 dc motor field and armature winding connections. (a) Separately excited connected; (b) shunt connected; (c) series connected; and (d) cumulative compound connected.

7.6 DIRECT-CURRENT MOTORS

EXAMPLE 7.15

The motor of Example 7.13 is shunt connected such that its field current is 1.0 A. Calculate its developed torque for the following operating conditions:

$$V_T = 80 \text{ V}$$
$$r_a = 0.2 \text{ }\Omega$$
$$\omega_m = 110 \text{ rad/s}$$

First, calculate E_g using the motor's magnetization curve:

$$E_{g100} = 71 \text{ V} \quad \text{for} \quad i_f = 1.0 \text{ A}$$

$$K_a\phi = \frac{E_g}{\omega_m} = 0.71$$

$$E_{g110} = K_a\phi\omega_m$$
$$= 0.71 \times 110 = 78.1 \text{ V}$$

Now calculate I_a:

$$I_a = \frac{V_T - E_{g110}}{r_a}$$
$$= \frac{80 - 78.1}{0.2} = 9.5 \text{ A}$$

The developed torque is

$$\tau_d = \frac{E_g I_a}{\omega_m}$$
$$= \frac{78.1 \times 9.5}{110} = 6.745 \text{ N·m} \quad \blacksquare\blacksquare$$

EXAMPLE 7.16

The motor of Example 7.13 is cumulative compound connected such that its armature current is 20 A. Calculate the shunt field current required to drive the motor at 95 rad/s. The motor characteristics are as follows:

$$V_T = 70 \text{ V}$$
$$r_a = 0.2 \text{ }\Omega$$
$$r_{se} = 0.1 \text{ }\Omega \quad \text{(series field resistance)}$$
$$N_{se}/N_{fe} = 0.01 \quad \text{(series field to shunt field turns ratio)}$$

As in the previous example, the relationship between the speed E_g and field excitation is the magnetization curve. However, in this case, the field excitation is provided by both the series and shunt windings. To use the magnetization curve of Figure 7.36, we can represent this combined excitation as an excitation due to the shunt field alone with an effective field current that accounts for the effects of the additional series excitations:

$$N_f I_f^* = N_f I_f + N_{se} I_{se}$$

or

$$I_f^* = I_f + \frac{N_{se}}{N_f} I_{se}$$

where I_f^*, the effective field current, is the quantity used on the magnetization curve.

To begin, calculate E_g:

$$\begin{aligned} E_g &= V_T - I_a(r_a + r_{se}) \\ &= 70 - (20)(0.3) \\ &= 64 \text{ V} \quad \text{at 95 rad/s} \end{aligned}$$

Use this result to calculate the equivalent E_g at 100 rad/s:

$$K_{a\phi} = \frac{E_{g100}}{100} = \frac{E_{g95}}{95}$$

$$E_{g100} = \frac{100}{95} \times E_{g95}$$

$$= \frac{100}{95} \times 64 = 67.37 \text{ V}$$

From the magnetization curve the effective field current corresponding to E_{g100} is 0.85 A. Substitute this quantity into the expression for I_f^* shown above and solve for I_f:

$$I_f^* = I_f + \frac{N_{se}}{N_f} I_{se}$$

$$I_f = I_f^* - \frac{N_{se}}{N_f} I_{se} = I_f^* - \frac{N_{se}}{N_f} I_a$$

$$= 0.85 - 0.01 \times 20 = 0.65 \text{ A} \qquad \blacksquare\blacksquare$$

7.7 UNIVERSAL MOTORS

One of the more popular motors used in single-phase applications is the universal motor. Its design is shown in Figure 7.39.

In this motor current flows into two windings connected in series. These windings are called the field and armature. The armature is wound on the rotating piece and receives its current through a set of brushes and commutator just as the armature of a dc motor does. The field is connected in series with the armature. This is the same design as a series-connected dc motor. In fact, most universal motors may be operated with direct or alternating current.

When universal motors operate with alternating current, the series connection of the field and armature windings keeps the field MMF and armature

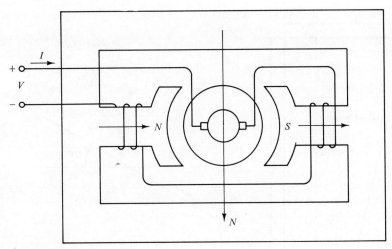

Figure 7.39 Universal motor.

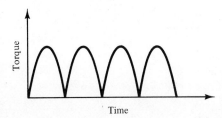

Figure 7.40 Torque variation in universal motors.

MMF in the same relative orientation. That is, when the current in the windings reverses direction, the two magnetic fields reverse their directions simultaneously, and the direction of developed torque on the rotor remains unchanged. However, the magnitude of the torque does not remain constant. The alternating current in the windings causes a variation in torque as shown in Figure 7.40. This feature is one of the disadvantages of universal motors because it means that they are noisy and experience unwanted vibration.

7.8 SUMMARY

This chapter has examined the major component of power system loads, that is, electric motors. The three kinds of motors presented were synchronous, induction, and direct current. The differences between them involve the manner in which they sustain rotation while minimizing pulsating torques. For synchronous and induction motors the rotational effort comes from the rotor MMF trying to align itself with a rotating air gap flux. For dc motors the flux and MMF are held stationary in relative positions that yield constant peak torque. The stationary rotor MMF is achieved with a commutator.

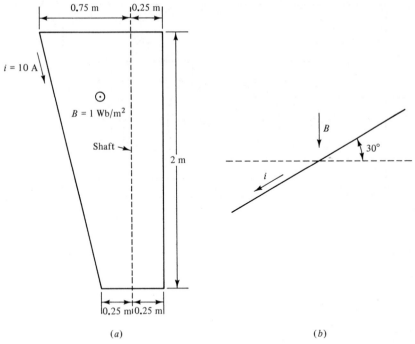

Figure 7.41 Coil; (a) front view and (b) top view.

7.9 PROBLEMS

7.1. Figure 7.41 shows a single-turn coil carrying 10 A of current suspended in a magnetic field. Calculate the following quantities:
  (a) forces acting on the left and right sides of the coil
  (b) torque that will cause rotation around the shaft

7.2. A 60-Hz, three-phase, four-pole synchronous motor is operating with a field excitation voltage of $230/\!-\!10°$ (l-n) while connected to a $220/0°$ (l-n) bus. Calculate the power input to the motor per phase from the bus and the torque applied to its shaft. The motor's synchronous reactance is 1 Ω.

7.3. Repeat Problem 7.2 for a field excitation voltage of $220/\!-\!10°$ (l-n).

7.4. A synchronous motor and inductive load are connected in parallel to an infinite bus as shown in Figure 7.42. Calculate the total power delivered by the infinite bus.

7.5. Calculate a new value of E_f shown in Figure 7.42 that will reduce the total reactive power delivered by the infinite bus to 0.

7.6. A three-phase, eight-pole, 60-Hz induction motor is operating at a speed of 855 rpm. Determine the slip at which the machine is operating and the frequency of the rotor currents and voltages.

7.7. A 30-hp, three-phase, 60-Hz, Y-connected, 220-V, six-pole, wound rotor induction motor has the following impedances in ohms referred to the stator:

$r_1 = 0.09 \qquad X_1 = 0.20 \qquad r_2 = 0.07 \qquad X_2 = 0.15 \qquad X_m = 8.00$

7.9 PROBLEMS

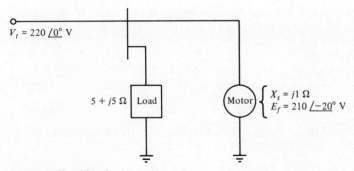

Figure 7.42 Circuit.

The machine is operated at rated voltage and frequency. Friction, windage, and core losses are 540 W. Determine the stator current, output torque, and efficiency of the motor at a slip of 3 percent.

7.8. For the motor in Problem 7.7 determine the starting torque and starting stator current.

7.9. Determine the maximum torque of the machine in Problem 7.7 and the slip at which it occurs.

7.10. What value of external resistance would have to be added to the rotor to increase the starting torque of the machine in Problem 7.7 by 50 percent?

7.11. For the machine in Problem 7.7 determine the copper losses and power across the air gap for operating slips of 2, 4, and 6 percent.

7.12. A single-phase induction motor has characteristics as shown below. Calculate the following quantities for its operation at a slip of 0.05.
  (a) line current
  (b) voltage induced on the main winding
  (c) total air gap, power
  (d) rotor copper loss
  (e) developed mechanical power
  (f) efficiency with respect to P_{out} where $P_{out} = P_d - $ (rotational losses)

$$r_1 = 2.0\,\Omega \quad r_2 = 2.5\,\Omega \quad X_m = 55\,\Omega$$
$$X_1 = 3.0\,\Omega \quad X_2 = 1.8\,\Omega \quad V_m = 115\underline{/0°}\text{ V}$$
$$\text{rotational losses} = 40\text{ W}$$

7.13. Repeat Problem 7.12 for the motor with characteristics as shown below. Use a slip of 0.02.

$$r_1 = 1.5\,\Omega \quad r_2 = 1.5\,\Omega \quad X_m = 65\,\Omega$$
$$X_1 = 2.0\,\Omega \quad X_2 = 2.0\,\Omega \quad V_m = 115\underline{/0°}\text{ V}$$
$$\text{rotational losses} = 40\text{ W}$$

7.14. A dc motor has a magnetization curve as shown in Figure 7.36. Calculate the voltage induced on its armature for a speed of 95 rad/s and field current of 0.5 A.

7.15. A dc motor has a magnetization curve as shown in Figure 7.36. It is operated at 105 rad/s with $K_a \phi$ of 0.5 V·s/rad. Calculate its field current and voltage induced on its armature.

7.16. The motor of Problem 7.14 has characteristics as shown below. If the motor is separately excited, calculate the following quantities:
(a) armature input power
(b) developed power
(c) developed torque
(d) efficiency for rotational losses of 100 W

$$V_T = 60 \text{ V} \qquad r_a = 0.1 \, \Omega$$

7.17. Repeat Problem 7.16 for the motor of Problem 7.15. Its characteristics are as follows:

$$V_T = 55 \text{ V} \qquad r_a = 0.2 \, \Omega$$

7.18. A shunt-connected dc motor has a magnetization curve as shown in Figure 7.36. Its characteristics are given below. Calculate its speed.

$$V_T = 70 \text{ V} \qquad r_a = 0.1 \, \Omega \qquad r_f = 80 \, \Omega \qquad \text{line current} = 30 \text{ A}$$

7.19. A cumulative compound-connected dc motor has a magnetization curve as shown in Figure 7.36. Its characteristics are given below. Calculate its speed.

$$V_T = 75 \text{ V} \qquad r_a = 0.2 \, \Omega \qquad r_{sc} = 0.05 \, \Omega \qquad r_f = 100 \, \Omega$$
$$N_{sc}/N_f = 0.01 \qquad \text{line current} = 40 \text{ A}$$

7.21. A shunt-connected dc motor has a magnetization curve as shown in Figure 7.36. Its characteristics are given below. Calculate its starting current.

$$V_T = 70 \text{ V} \qquad N_a = 0.1 \, \Omega \qquad r_f = 80 \, \Omega$$

7.22. Calculate the additional armature circuit resistance needed to limit the starting current to 100 A in the motor of Problem 7.21.

Appendix A

Phasor Analysis

A.1 PHASOR REPRESENTATION OF SINUSOIDS

Sinusoidal varying quantities, such as voltages and currents, can be represented at any instant in time by equations of the form

$$i(t) = I_m \cos(\omega t + \phi) \quad \text{A} \tag{A.1}$$

and
$$v(t) = V_m \cos(\omega t) \quad \text{V} \tag{A.2}$$

where I_m = the maximum value of $i(t)$, A
 $\omega = 2\pi f$ = angular frequency, rad/s
 ϕ = phase angle, rad
 V_m = the maximum value of $v(t)$, V

The graphs of Equations A.1 and A.2 are shown in Figure A.1. The values of $v(t)$ and $i(t)$ at some time t can also be determined graphically. If vectors of magnitude I_m

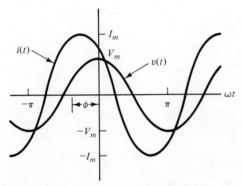

Figure A.1 Graph of $i(t)$ and $v(t)$.

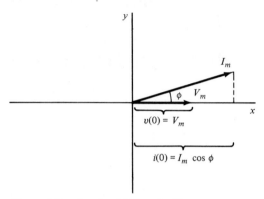

Figure A.2 Graph of $i(0)$ and $v(0)$.

and V_m are placed on a graph and displaced from the x axis by an angle equaling the phase angles of $i(t)$ and $v(t)$, respectively, the projection of these vectors onto the x axis will be the value of $i(t)$ and $v(t)$ at time $t = 0$. This concept is shown in Figure A.2.

If these vectors rotate counterclockwise at an angular velocity of ω, then at any instant in time the projection of I_m and V_m onto the x axis will be the value of $i(t)$ and $v(t)$, respectively. Therefore, if all voltages and currents of a circuit have the same frequency, the circuit can be solved at a particular instant, such as depicted in Figure A.2, and the solution of the circuit variables will be known for all times. The analysis of these rotating vectors at a particular time (similar to taking a photograph of the rotating vectors) is called phasor analysis. I_m and V_m in Figure A.2 are phasors.

EXAMPLE A.1

Two voltage sources, which are to be connected in series, are generating the following voltages:

$$v_1(t) = 100\cos(377t) \quad \text{V}$$
$$v_2(t) = 150\cos(377t + \pi/6) \quad \text{V}$$

Find the voltage across the series combination using phasor analysis. At time $t = 0$ the phasor diagram looks like this:

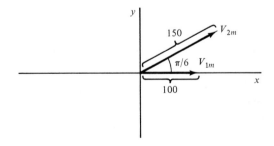

The notations for these two phasors are

$$V_{1m} = 100\underline{/0°} \qquad V_{2m} = 150\underline{/30°}$$

A.1 PHASOR REPRESENTATION OF SINUSOIDS

Since the two sources are in series, their voltages will sum:

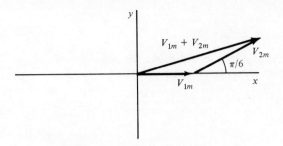

Mathematically, this operation is done by summing the projections onto the x and y axes of V_{1m} and V_{2m}:

$$V_{1m} + V_{2m}|_x = V_{1m}\cos(0) + V_{2m}\cos(\pi/6) = 230$$
$$V_{1m} + V_{2m}|_y = V_{1m}\sin(0) + V_{2m}\sin(\pi/6) = 75$$

Therefore, the summation of $V_{1m} + V_{2m}$ is

$$V_{Tm} = V_{1m} + V_{2m} = 230 + j75 = 242\underline{/18°} \text{ V}$$

The time-varying value of $V_{1m} + V_{2m}$ is

$$v_T(t) = v_1(t) + v_2(t) = 242\cos(377t + \pi/10) \quad \text{V} \qquad \blacksquare\blacksquare$$

Note that the phasor solution V_{Tm} was written in two forms:

$$V_{Tm} = 230 + j75 = 242\underline{/18°} \text{ V} \tag{A.3}$$

The polar form indicates that the magnitude of the phasor V_{Tm} is 242 V, and for the instant at which the analysis was done it formed an angle of 18° or $\pi/10$ rad with the x axis. The cartesian form says that V_{Tm} is composed of 230 V at an angle of 0° and 75 V shifted 90° from the x axis (the j operator shifts all quantities by 90°). From complex number analysis it is seen that these two forms are equal:

$$\sqrt{230^2 + 75^2} = 242 \tag{A.4}$$

$$\tan^{-1}\left(\frac{75}{230}\right) = 18° \tag{A.5}$$

Since the phasor representation exactly parallels complex number analysis, the projection of a phasor onto the x axis is called the real part, and the projection onto the y axis is called the imaginary part. It is interesting to note that the real part of a phasor is the actual instantaneous value of the quantity at the particular time of analysis.

In Appendix B it is shown that in order to calculate average power from sinusoidal voltages and currents, it is necessary to use rms values of voltages and currents. Since average real and reactive powers are of primary concern to power engineers, voltages and currents are usually expressed as rms values. Therefore, in the power industry, phasor representations are written

$$\text{voltage} = V\underline{/\theta} \tag{A.6}$$

$$\text{current} = I\underline{/\alpha} \tag{A.7}$$

where V and I are rms values of the sinusoidal voltages and currents. If V_m is the maximum value of the sinusoidal voltage represented by the phasor in Equation A.6, then

$$V = V_m/\sqrt{2} \tag{A.8}$$

In this text, as well as most power engineering related books, the magnitude of voltage and current phasors are given as rms values unless otherwise stated.

A.2 IMPEDANCES

From circuit analysis the relationship between time-varying voltages and currents applied to resistors, inductors, and capacitors are

$$v(t) = i(t)r \tag{A.9}$$

$$v(t) = L\frac{di(t)}{dt} \tag{A.10}$$

$$i(t) = C\frac{dv(t)}{dt} \tag{A.11}$$

If $i(t)$ in Equation A.9 is sinusoidal and has the value

$$i(t) = I_m \sin(\omega t + \phi) \quad \text{A} \tag{A.12}$$

then Equation A.9 becomes

$$v(t) = rI_m \sin(\omega t + \phi) = V_m \sin(\omega t + \phi) \quad \text{V} \tag{A.13}$$

where V_m is the maximum value of $v(t)$ and equals rI_m. In phasor notation Equation A.13 can be written

$$V\underline{/\alpha} = rI\underline{/\phi} \quad \text{V} \tag{A.14}$$

Note that since it is not stated otherwise V and I are rms values. Also since r is a scalar, there is no phase shift and $\alpha = \phi$.

Using the current in Equation A.12, Equation A.10 can be written as

$$v(t) = L\frac{dI_m \sin(\omega t + \phi)}{dt}$$

$$= \omega L I_m \cos(\omega t + \phi)$$
$$= \omega L I_m \sin(\omega t + \phi + \pi/2)$$
$$= V_m \sin(\omega t + \phi + \pi/2) \quad \text{V} \tag{A.15}$$

where V_m is the maximum value of $v(t)$ and equals $\omega L I_m$. In phasor notation Equation A.15 becomes

$$V\underline{/\alpha} = \omega L I\underline{/\phi + \pi/2} = j\omega L I\underline{/\phi} \quad \text{V} \tag{A.16}$$

Since the j operator shifts phasors by 90°, the voltage of Equation A.16 leads the current by $\pi/2$ rad. Therefore, the phase angle of the voltage is

$$\alpha = \phi + 90° \tag{A.17}$$

A.2 IMPEDANCES

If the same analysis is done for Equation A.11, the phasor equation is found to be

$$I\underline{/\phi} = j\omega C V \underline{/\alpha} \quad \text{A} \tag{A.18}$$

In this equation the current leads the voltage by 90°.

Equations A.14, A.16, and A.18 can be written in a standard form using impedance instead of resistance, inductance, and capacitance:

$$V = IZ \tag{A.19}$$

Impedance is a complex number and is directly related to the values of the circuit elements. The values of impedance that correspond to values of resistance, inductance, and capacitance are, respectively,

$$Z = r + j0 \tag{A.20}$$

$$Z = 0 + j\omega L \tag{A.21}$$

$$Z = 0 - \frac{j}{\omega C} \tag{A.22}$$

Impedance has the units of ohms and combines in series and parallel similarly to resistance. Two impedances, Z_1 and Z_2, in series add to give a total impedance:

$$Z_T = Z_1 + Z_2 \tag{A.23}$$

In parallel these two same impedances would combine as follows:

$$Z_T = \frac{1}{1/Z_1 + 1/Z_2} = \frac{Z_1 Z_2}{Z_1 + Z_2} \tag{A.24}$$

The real part of an impedance is called resistance and is represented by r. The imaginary part is called reactance and is represented by X.

EXAMPLE A.2

An 8.66-Ω resistor and a 0.0133-H inductor are connected in series. A 60-Hz voltage of 240 V is placed across the combination. Using phasor analysis, find the value of current flowing through the series connection.

Since it is not stated otherwise, the value of 240 V is an rms value. We shall choose the voltage as reference. This means that the phase angle of the voltage is assumed to be 0°. All this assumption does is choose the instant in time at which we do our analysis.

$$V = 240\underline{/0°} \text{ V}$$

The impedance of the series connection is

$$Z = 8.66 + j(377)0.0133 = 8.66 + j5 = 10\underline{/30°} \text{ Ω}$$

Therefore, the current flowing through the impedance is

$$I = \frac{240\underline{/0°}}{10\underline{/30°}} = 24\underline{/-30°} \text{ A}$$ ■■

Appendix B

Average Power and Three-Phase Calculations

B.1 AVERAGE POWER

Instantaneous power p is defined as instantaneous voltage v multiplied by instantaneous current i:

$$p = vi \quad \text{W} \tag{B.1}$$

If voltage and current are cyclic functions, then an average power over the period of the cycle can be calculated:

$$P = \frac{1}{T}\int_0^T p(t)\,dt \quad \text{W} \tag{B.2}$$

where P = average power
T = period of cycle

If the voltage and current are sinusoidal functions and can be represented by

$$v(t) = V_m \cos \omega t \quad \text{V} \tag{B.3}$$

$$i(t) = I_m \cos(\omega t - \theta) \quad \text{A} \tag{B.4}$$

then Equation B.1 becomes

$$p(t) = V_m I_m \cos \omega t \cos(\omega t - \theta) \quad \text{W} \tag{B.5}$$

where V_m and I_m are maximum values of voltage and current, respectively. Equation B.5 can be expanded to

$$\begin{aligned}p(t) &= V_m I_m \tfrac{1}{2}[\cos(\omega t - \omega t + \theta) + \cos(\omega t + \omega t - \theta)] \\ &= \tfrac{1}{2} V_m I_m \cos \theta + \tfrac{1}{2} V_m I_m \cos(2\omega t - \theta)\end{aligned} \tag{B.6}$$

Applying the integration of Equation B.2 to Equation B.6, we find that the average of a time-varying sinusoidal function is 0. Therefore, integration of the last term of

B.2 POWER FACTOR AND COMPLEX POWER

Equation B.6 gives 0, and we are left with the constant term

$$P = \tfrac{1}{2} V_m I_m \cos \theta \quad \text{W} \tag{B.7}$$

If the $\tfrac{1}{2}$ term is divided between the voltage and current maximums, then Equation B.7 becomes

$$P = \frac{V_m}{\sqrt{2}} \frac{I_m}{\sqrt{2}} \cos \theta$$

$$= VI \cos \theta \quad \text{W} \tag{B.8}$$

where V and I are the rms or effective values of voltage and current. These values are the magnitudes of the phasors which would represent the voltage and current of Equations B.3 and B.4.

Effective values will be defined using the current in Equation B.4:

$$I = \left(\frac{1}{T} \int_0^T i^2 \, dt \right)^{1/2} \quad \text{A} \tag{B.9}$$

Notice that the rms value is the square root of the average of the square of the instantaneous current. This calculation yields

$$I = \left[\frac{1}{T} \int_0^T I_m^2 \cos^2(\omega t - \theta) \, dt \right]^{1/2}$$

$$= \left\{ \frac{\omega}{2\pi} \int_0^{2\pi/\omega} I_m^2 [\tfrac{1}{2} + \tfrac{1}{2} \cos(2\omega t - 2\theta)] \, dt \right\}^{1/2} \quad \text{A} \tag{B.10}$$

where $T = 1/f = 2\pi/\omega$.

Since the average of a time-varying sinusoidal function over one period is 0, Equation B.10 reduces to

$$I = I_m \left(\frac{\omega}{2\pi} \int_0^{2\pi/\omega} \tfrac{1}{2} \, dt \right)^{1/2}$$

$$= I_m \left[\frac{\omega}{2\pi} \left(\frac{1}{2} \right) \frac{2\pi}{\omega} \right]^{1/2}$$

$$= I_m / \sqrt{2} \quad \text{A} \tag{B.11}$$

The rms value of any sinusoidal function is the maximum of that function divided by $\sqrt{2}$.

B.2 POWER FACTOR AND COMPLEX POWER

From Equation B.8 it is seen that average power is a function not only of rms voltage and current but also of the phase angle difference between the two. If the voltage and current of Equations B.3 and B.4 are in phase, then θ is equal to $0°$. This means that Equation B.8 becomes

$$P = VI \cos 0 = VI \quad \text{W} \tag{B.12}$$

Equations B.13 and B.14 show the calculations of average power for phase angle differences of $60°$ and $90°$, respectively:

$$P = VI \cos(\pi/3) = VI(0.5) \tag{B.13}$$

$$P = VI\cos(\pi/2) = 0 \tag{B.14}$$

Clearly, for the same magnitude of voltage and current, the corresponding magnitude of average power can range from zero to VI.

It is of interest to calculate the average power dissipated in various passive elements. The current through a resistor is in phase with the voltage across it:

$$V_r = I_r r \tag{B.15}$$

where V_r and I_r are phasor quantities. Therefore, the average power dissipated in a resistor is simply the voltage across it multiplied by the current through it:

$$\begin{aligned} P &= V_r I_r \cos 0 \\ &= VI \end{aligned} \tag{B.16}$$

For inductors and capacitors the current through these elements lags and leads, respectively, the voltage across them by 90°:

$$V_L = I_L j\omega L \tag{B.17}$$

$$V_c = I_c\left(\frac{-j}{\omega C}\right) \tag{B.18}$$

where V_L, V_c, I_L, and I_c are phasor quantities. Therefore, no average power is dissipated in these elements.

The magnitude of the voltage multiplied by the magnitude of the current is given the name apparent power. The ratio of average power to apparent power is called the power factor. For sinusoidal voltages and currents this is simply $\cos\theta$:

$$\begin{aligned} \text{pf} &= \frac{P}{VI} = \frac{VI\cos\theta}{VI} \\ &= \cos\theta \end{aligned} \tag{B.19}$$

θ is called the power factor angle and is the angle by which the voltage leads the current.

When a voltage is placed across a load, the impedance of that load determines the magnitude and angle of the current that will flow through the load. That is, power factor is a characteristic of the load. Therefore, if two loads required the same power and were supplied by the same voltage source, but load 1 had a pf of 1 and load 2 had a pf of 0.5, then twice as much current would be required to deliver the same power to load 2 as would be required for load 1. For efficient and economic operation, power factors close to 1 are usually desirable.

It is sometimes convenient to consider average power as the real part of a quantity called complex power. Complex power is defined as

$$S = V_a I_a^* \quad \text{VA} \tag{B.20}$$

where S is a complex number, V_a and I_a are phasor quantities, and I_a^* is the conjugate of I_a. If V_a and I_a are defined as

$$V_a = V\underline{/\theta_1} \tag{B.21}$$

$$I_a = I\underline{/\theta_2} \tag{B.22}$$

then Equation B.20 becomes

$$S = VI\cos(\theta_1 - \theta_2) + jVI\sin(\theta_1 - \theta_2) \tag{B.23}$$

Since $\theta_1 - \theta_2$ is the angle by which the voltage leads the current, the real part of S is by definition average power. Because of this, average power is sometimes called real power. Average power is also often simply called power. From Equation B.23 it is obvious that the magnitude of S is equal to apparent power. Apparent power is, therefore, given the symbol $|S|$.

The imaginary part of S is called reactive power and is given the symbol Q and the units vars. Just as real power can only be dissipated in a resistance, reactive power can only be dissipated in a reactance. Notice that Equation B.20 will result in positive reactive power being dissipated in an inductive reactance since current lags voltage for an inductor. By the same reasoning, negative reactive power will be dissipated in a capacitor. This means that capacitors will supply reactive power when a voltage is placed across them.

Since power factor is a cosine function and has the same value for positive or negative angles, the word *lag* or *lead* should be written after the power factor to more fully describe the load. If the power factor is followed by the word *lag*, the current lags the voltage and the load is inductive. Similarly, if it is followed by *lead*, the load is capacitive.

B.3 THREE-PHASE CALCULATIONS

Almost the entire output of the electric utility industry in this country is generated, transmitted, and distributed on a three-phase system. At the generating stations three voltages are generated. They are of equal magnitude but each is 120° out of phase with the other two. This is called a balanced source. Each of these source voltages is connected by a separate conductor to different loads. If all three voltages are used at a load, the load is called a three-phase load. If each voltage supplies the same magnitude and type of load, the load is said to be balanced. When a balanced source supplies a balanced load, then in each phase of the load and in each conductor currents will flow that are equal in magnitude and 120° out of phase. These currents are said to be balanced currents. Figure B.1 shows a simple circuit and phasor diagram for a balanced system.

The system shown in Figure B.1 is said to be an *abc* sequence since phase *b* lags phase *a* by 120° and phase *c* lags phase *b* by 120°. The only other possible sequence would be an *acb* sequence.

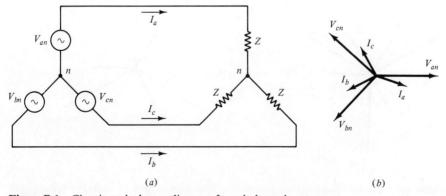

Figure B.1 Circuit and phasor diagram for a balanced system.

The load in Figure B.1 is connected in a Y connection. With this type of connection the voltage across each phase of the load is a line-to-neutral voltage, and the current flowing in each phase of the load is the current flowing in each line or the line current. The voltages between each line in Figure B.1 can be easily calculated:

$$V_{ab} = V_{an} + V_{nb} = V_{an} - V_{bn}$$
$$V_{bc} = V_{bn} - V_{cn} \quad (B.24)$$
$$V_{ca} = V_{cn} - V_{an}$$

Performing the indicated mathematical operation, we find that for the *abc* sequence of Figure B.1

$$V_{ab} = V_{an}\sqrt{3}\underline{/30°}$$
$$V_{bc} = V_{bn}\sqrt{3}\underline{/30°} \quad (B.25)$$
$$V_{ca} = V_{cn}\sqrt{3}\underline{/30°}$$

Each line-to-line voltage leads its corresponding line-to-neutral voltage by 30° and is the $\sqrt{3}$ times its magnitude. For an *acb* sequence Equations B.25 is modified to

$$V_{ab} = V_{an}\sqrt{3}\underline{/-30°}$$
$$V_{bc} = V_{bn}\sqrt{3}\underline{/-30°} \quad (B.26)$$
$$V_{ca} = V_{cn}\sqrt{3}\underline{/-30°}$$

For this sequence the line-to-line voltages lag the line-to-neutral voltages by 30°. Phasor diagrams of these relationships are shown in Figure B.2.

The power being consumed in each phase of the load in Figure B.1 is

$$P_{1\phi} = |V_{an}|I_l \cos \theta \quad (B.27)$$

where I_l is the magnitude of I_a and $\cos \theta$ is the power factor associated with each phase of the load. Since the system is balanced, the total power being consumed is

$$P_T = 3P_{1\phi} = 3|V_{an}|I_l \cos \theta$$
$$= 3 \frac{V_{ll}}{\sqrt{3}} I_l \cos \theta$$
$$= \sqrt{3} V_{ll} I_l \cos \theta \quad (B.28)$$

where V_{ll} is the magnitude of the line-to-line voltage.

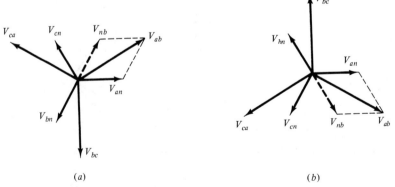

(a)   (b)

Figure B.2 Phasor diagrams of line-to-line and line-to-neutral voltages for (a) *abc* sequence and (b) *acb* sequence.

B.3 THREE-PHASE CALCULATIONS

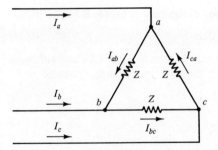

Figure B.3 Δ connection.

The load in Figure B.1 is Y-connected. There is one other possible connection for balanced three-phase loads. This connection is called a Δ connection and is illustrated in Figure B.3. The voltage across each phase of the load for this connection is the line-to-line voltage. The relationship between the currents flowing in each phase of the load and the line currents are given in Equations B.29 and Figure B.4.

$$I_a = I_{ab} + I_{ac} = I_{ab} - I_{ca}$$
$$I_b = I_{bc} - I_{ab} \quad \quad (B.29)$$
$$I_c = I_{ca} - I_{bc}$$

The relationships between line currents and Δ phase currents for *abc* and *acb* sequences are given in Equations B.30 and B.31, respectively:

$$I_a = I_{ab}\sqrt{3}\underline{/-30°}$$
$$I_b = I_{bc}\sqrt{3}\underline{/-30°} \quad \quad (B.30)$$
$$I_c = I_{ca}\sqrt{3}\underline{/-30°}$$

For an *abc* sequence the line currents are $\sqrt{3}$ times larger than the Δ phase currents and lag them by 30°:

$$I_a = I_{ab}\sqrt{3}\underline{/30°}$$
$$I_b = I_{bc}\sqrt{3}\underline{/30°} \quad \quad (B.31)$$
$$I_c = I_{ca}\sqrt{3}\underline{/30°}$$

For an *acb* sequence the line currents lead the Δ phase currents by 30°.

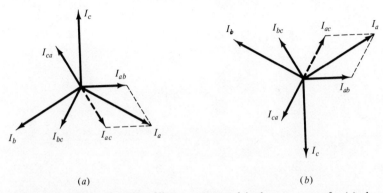

Figure B.4 Phase diagram of line currents and Δ phase currents for (*a*) *abc* sequence and (*b*) *acb* sequence.

The power being consumed in each phase of the load in Figure B.3 is

$$P_{1\phi} = V_{ll}|I_{ab}|\cos\theta \tag{B.32}$$

where V_{ll} is the magnitude of V_{ab} and $\cos\theta$ is the power factor associated with each phase of the load. Since the system is balanced, the total power being consumed is

$$P_T = 3P_{1\phi} = 3V_{ll}|I_{ab}|\cos\theta$$

$$= 3V_{ll}\frac{I_l}{\sqrt{3}}\cos\theta$$

$$= \sqrt{3}V_{ll}I_l\cos\theta \tag{B.33}$$

where I_l is the magnitude of the line current.

Note that Equation B.33 is exactly the same as Equation B.28. This means that if the line-to-line voltage and line current are known at a balanced load and the power factor of that load is known, the power being consumed by that load can be calculated without knowing how the load is connected.

Similar derivations can be done to show that

$$|S_T| = \sqrt{3}V_{ll}I_l \tag{B.34}$$

and

$$Q_T = \sqrt{3}V_{ll}I_l\sin\theta \tag{B.35}$$

where θ is the angle by which the voltage across each phase of the load leads the current flowing through each phase of the load.

Appendix C

Matrix Algebra

C.1 DEFINITIONS

The arrangement of numbers of functions in an orderly array is called a matrix. Matrix A in Equation C.1 has the order m by n:

$$A = \begin{bmatrix} a_{11} & a_{12} & \cdots & a_{1n} \\ a_{21} & a_{22} & \cdots & a_{2n} \\ \vdots & \vdots & \vdots & \vdots \\ a_{m1} & a_{m2} & \cdots & a_{mn} \end{bmatrix} \quad (C.1)$$

A matrix which has the same number of rows as columns is called a square matrix. The elements $a_{11}, a_{22}, \ldots,$ of a square matrix are called the diagonal elements. All other elements are known as off-diagonal elements. A square matrix that has 1 as all the diagonal elements and 0 as all the off-diagonal elements is called the unit matrix U.

C.2 ALGEBRAIC OPERATIONS

Two matrices of the same order can be added together. This is done by summing the corresponding elements of each matrix:

$$C = A + B = \begin{bmatrix} a_{11} & a_{12} \\ a_{21} & a_{22} \\ a_{31} & a_{32} \end{bmatrix} + \begin{bmatrix} b_{11} & b_{12} \\ b_{21} & b_{22} \\ b_{31} & b_{32} \end{bmatrix}$$

$$= \begin{bmatrix} a_{11}+b_{11} & a_{12}+b_{12} \\ a_{21}+b_{21} & a_{22}+b_{22} \\ a_{31}+b_{31} & a_{32}+b_{32} \end{bmatrix} = \begin{bmatrix} c_{11} & c_{12} \\ c_{21} & c_{22} \\ c_{31} & c_{32} \end{bmatrix} \quad (C.2)$$

Note that
$$A + B = B + A \tag{C.3}$$

In order to multiply two matrices, A and B, B must have the same number of rows as A has columns. If A is 2 by 3 and B is 3 by 1 then C of Equation C.4 will be 2 by 1 (the number of rows of A and columns of B).

$$C = AB \tag{C.4}$$

The formula for each element of C would be

$$c_{ij} = \sum_{k=1}^{r} a_{ik} b_{kj} \tag{C.5}$$

where r is the number of columns of A and rows of B:

$$\begin{bmatrix} a_{11} & a_{12} & a_{13} \\ a_{21} & a_{22} & a_{23} \end{bmatrix} \begin{bmatrix} b_{11} \\ b_{21} \\ b_{31} \end{bmatrix} = \begin{bmatrix} a_{11}b_{11} + a_{12}b_{21} + a_{13}b_{31} \\ a_{21}b_{11} + a_{22}b_{21} + a_{23}b_{31} \end{bmatrix} = \begin{bmatrix} C_{11} \\ C_{21} \end{bmatrix} \tag{C.6}$$

In general, even if both AB and BA are defined,

$$AB \neq BA \tag{C.7}$$

The unit matrix is, however, an exception to this rule when multiplied by a square matrix A:

$$UA = AU = A \tag{C.8}$$

Any matrix may be multiplied by a scalar by multiplying every element in the matrix by the scalar.

C.3 INVERSE OF A MATRIX

Division is not defined for matrices. However, the algebraic objective of division can be obtained by introducing the concept of inverse of a matrix. Only square matrices may be inverted. The inverse of a matrix times the original matrix equals the unit matrix.

$$A^{-1}A = AA^{-1} = U \tag{C.9}$$

where A is a square matrix, and A^{-1} is the inverse of A.

The inverse operation can be very useful. If, for instance, the values of matrix A and C are known and A is a square matrix, the values of matrix B in Equation C.10 can be determined:

$$C = AB \tag{C.10}$$

Premultiplying both sides by A^{-1},

$$A^{-1}C = A^{-1}AB = UB = B \tag{C.11}$$

C.4 MATRIX PARTITIONING

Matrices can be divided into submatrices, and each of these submatrices can then be treated as elements of the original matrices. As an example, the matrices in Equations

C.4 MATRIX PARTITIONING

C.12 and C.13 can be subdivided as shown:

$$A = \begin{bmatrix} a_{11} & a_{12} & a_{13} \\ a_{21} & a_{22} & a_{23} \\ \hline a_{31} & a_{32} & a_{33} \end{bmatrix} = \begin{bmatrix} D & E \\ \hline F & G \end{bmatrix} \quad \text{(C.12)}$$

$$B = \begin{bmatrix} b_{11} \\ b_{21} \\ \hline b_{31} \end{bmatrix} = \begin{bmatrix} H \\ \hline J \end{bmatrix} \quad \text{(C.13)}$$

The multiplication of these two matrices can now be represented as

$$AB = \begin{bmatrix} DH + EJ \\ FH + GJ \end{bmatrix} \quad \text{(C.14)}$$

Note that partitioning after the ith row requires partitioning after the ith column in order to keep the orders of the submatrices compatible.

Bibliography

Chapter 1

El-Hawary, M. E. and G. S. Christensen. *Optimal Economic Operation of Electric Power System.* New York: Academic Press, Inc., 1979.

"Electric Power Supply and Demand 1981–1990," National Electric Reliability Council, July 1981.

Fischetti, Mark A. "EPRI's First Decade." *IEEE Spectrum,* April 1983.

Kaplan, Gadi. "Prospects for Nuclear Power." *IEEE Spectrum,* March 1983.

Sullivan, R. L. *Power System Planning.* New York: McGraw-Hill Book Co., 1977.

Chapter 2

Durney, Carl H. and Curtis C. Johnson. *Introduction to Modern Electromagnetics.* New York: McGraw-Hill Book Co., 1969.

Mablekos, Van E. *Electric Machine Theory for Power Engineers.* New York: Harper & Row, Publishers, 1980.

Skitek, G. G. and S. V. Marshall. *Electromagnetic Concepts and Applications.* Englewood Cliffs, NJ: Prentice-Hall, Inc., 1982.

Slemon, G. R. and A. Straughen. *Electric Machines.* Reading, MA: Addison-Wesley Publishing Co., 1980.

Woodson, Herbert H. and James R. Melcher. *Electromechanical Dynamics, Part I: Discrete Systems.* New York: John Wiley & Sons, Inc., 1968.

Chapter 3

Elgerd, Olle I. *Electric Energy Systems Theory: An Introduction.* New York: McGraw-Hill Book Co., 1971.

BIBLIOGRAPHY

Fitzgerald, A. E., C. Kingsley, and A. Kusko. *Electric Machinery*, 3rd Ed. New York: McGraw-Hill Book Co., 1971.

Kimbark, Edward W. *Power System Stability: Synchronous Machines.* New York: Dover Publications, Inc., 1956.

McPherson, George. *An Introduction to Electric Machines and Transformers.* New York: John Wiley & Sons, Inc., 1981.

Woodson, Herbert H. and James R. Melcher. *Electromechanical Dynamics, Part I: Discrete Systems.* New York: John Wiley & Sons, Inc., 1968.

Chapter 4

Fitzgerald, A. E., C. Kingsley, and A. Kusko. *Electric Machinery*, 3rd Ed. New York: McGraw-Hill Book Co., 1971.

McPherson, George. *An Introduction to Electric Machines and Transformers.* New York: John Wiley & Sons, Inc., 1981.

Neuenswander, John R. *Modern Power Systems.* Scranton, PA: International Textbook Co., 1971.

Stevenson, W. D., Jr. *Elements of Power System Analysis*, 4th Ed. New York: McGraw-Hill Book Co., 1982.

Chapter 5

Elgerd, Olle I. *Electric Energy Systems Theory: An Introduction.* New York: McGraw-Hill Book Co., 1971.

Stevenson, W. D., Jr. *Elements of Power System Analysis*, 4th Ed. New York: McGraw-Hill Book Co., 1982.

Weeks, Walter L. *Transmission and Distribution of Electrical Energy.* New York: Harper & Row, Publishers, Inc., New York, 1981.

Westinghouse Electric Corp. *Electrical Transmission and Distribution Reference Book*, 4th Ed. East Pittsburgh, PA: Westinghouse Electric Corp., 1964.

Chapter 6

Elgerd, Olle I. *Electric Energy Systems Theory: An Introduction.* New York: McGraw-Hill Book Co., 1971.

Gross, C. A. *Power System Analysis.* New York: John Wiley & Sons, Inc., 1979.

Neuenswander, John R. *Modern Power Systems.* Scranton, PA: International Textbook Co., 1971.

Stagg, G. W. and A. H. El-Abiad. *Computer Methods in Power System Analysis.* New York: McGraw-Hill Book Co., 1968.

Stevenson, W. D., Jr. *Elements of Power System Analysis*, 4th Ed. New York: McGraw-Hill Book Co., 1982.

Chapter 7

Bodine Electric Co. *Small Motor, Gearmotor, and Control*, 4th Ed. Chicago: Bodine Electric Co., 1978.

Fitzgerald, A. E., C. Kingsley, and A. Kusko. *Electric Machinery*, 3rd Ed. New York: McGraw-Hill Book Co., 1971.

Mablekos, Van E. *Electric Machine Theory for Power Engineers*. New York: Harper & Row, Publishers, 1980.

McCormick, W. W. *Fundamentals of University Physics*. London: The Macmillan Co. Collier-Macmillan Limited, 1969.

McPherson, George. *An Introduction to Electric Machines and Transformers*. New York: John Wiley & Sons, Inc., 1981.

Slemon, G. R. and A. Straughen. *Electric Machines*. Reading, MA: Addison-Wesley Publishing Co., 1980.

Woodson, Herbert H. and James R. Melcher. *Electromechanical Dynamics, Part I: Discrete Systems*. New York: John Wiley & Sons, Inc., 1968.

Index

ac circuits
  balanced three-phase, 251
  power in balanced three-phase, 252–254
  power in single-phase, 252
Admittance
  mutual, 175
  self, 175
Admittance matrix, 173–176
Ampere's law, 14, 19, 128
Apparent power, 251
Armature reaction, 73
Autotransformer, 108, 109
Auxiliary winding, 222
Average power, 248, 249

Base for per unit quantities
  change of, 164–165
  current, 163
  impedance, 163
  power, 158–164, 165–170
  voltage, 158–164, 165–170
Boiler feed pumps, 49
Boilers, 48
Brushes, 55, 229
Bundled conductors, 122
Bus admittance matrix, 173–176

Capacitance
  calculation by method of images, 140
  definition, 137
  effect of earth on, 139
  effect of bundled conductors, 145, 146
  to neutral, 142
Capacitors
  as generators of reactive power, 187
Characteristic impedance, 148
Charge, 137, 138
Coenergy, 41
Commutator, 229
Complex power, 250
Condenser, 49
Conductance, 125
Conservation of charge, 15
Conservation of energy, 36
Contour integration, 17, 37
Control of P and Q
  by capacitors, 187
  by generators, 184–186
  by transformers, 187–189
Core losses, 96, 97
Corona, 122
Coulomb's law, 18
Current
  line, 251
  phase, 253
Current density, 20

dc motors
  armature, 232

dc motors (*Continued*)
  armature constant, 232
  brushes, 229
  circuit model, 234, 236
  commutation, 231, 232
  commutator, 229
  connections, 236
  elementary motor, 196
  magnetizing curve, 233
  power, 234
  structure, 229
  torque, 235
  voltage, 232–234
Δ connection, 253
Δ-Y transformers
  per unit impedance of, 169–172
  phase shift in, 114
Direct-axis synchronous inductance, 75

Economic dispatch, 5
Eddy currents, 96
Effective (rms) value, 249
Electric field
  around conductors, 138
  between earth and conductors, 139
  flux, 138
  intensity, 15
Electric power engineering, 6
Elementary generator, 52
Elementary motor, 193–197
Energy
  conservation, 3
  consumption growth, 3
Equivalent circuit
  of three-phase induction motor, 205–209
  of transformers, 103–105
  of transmission lines, 146–150

Faraday's law, 15, 17, 52
Faults, definition, 5
Field winding, 55, 58, 59
Flux linkages
  of coils, 52, 195
  internal, 128
  of isolated conductors, 127–130
  of one conductor in a group, 132–137
  varying, 53
Flux wave in induction motors, 201, 202
Force-coenergy relations, 42
Force-energy relations, 40, 66
Force of electric origin, 36

Gauss' law, 15
Gauss-Seidel method of load flow studies, 178–181, 182–184
Generator, 52
Geometric mean radius, 135–137

Harmonics, 110, 111
Hysteresis losses, 97

Inductance
  calculation, 66, 127–130
  effect of bundled conductors, 145
  single-phase transmission lines, 131, 132
  three-phase lines, 132–135
Induction motors
  breakdown torque, 205
  elementary motor, 193–196
  equivalent circuits, 205–209
  flux waves, 201, 202
  losses, 209, 210
  magnetizing current, 206
  magnetizing impedance, 206
  maximum torque, 216, 217
  referring to rotor quantities, 208–209
  rotor resistance, effect of, 218, 219
  single-phase, 219–228
  slip, 203
  squirrel cage rotor, 200
  stall, 205
  starting current, 213
  starting torque, 214
  steady-state analysis, 209–213
  Thevenin equivalent circuit, 214–216
  three-phase, 200–219
  torque, 210, 211, 217, 221
  torque-slip characteristics, 204
  wound rotor, 200
Infinite bus, 185
Instantaneous power, 248

j operator, 246

Lagging power factor, 251
Laminated cores, 97
Leading power factor, 251
Leakage inductance, 74
Leakage reactance, 93–95
Line-to-line voltages, 252
Line-to-neutral voltages, 252
Load bus, 176–178

INDEX **263**

Load characteristics, 3–5
Load flow
　data for, 182–184
　Gauss-Seidel method, 178–181, 182–184
　information obtained, 184
　nonconvergence, 180
　objectives of, 173
Loads, 193
Lossless systems, 36

Magnetic field
　coenergy, 41
　density, 14
　energy, 36
　flux, 13
　intensity, 14
Magnetic poles, 71
Magnetic source law, 15, 21
Magnetizing current
　in induction motors, 206
　in transformers, 96
Magnetomotive force, 23
Matrix
　addition, 255, 256
　inversion, 256
　multiplication, 256
　partitioning, 256, 257
　square, 255
　unit matrix, 256
Maxwell's equations, 14, 15
Multiwinding transformers, 106
Mutual admittance, 175

Off-nominal turns ratio transformers, 187–189
One-line diagrams, 156–158
Open-circuit test of transformers, 98

Per unit
　analysis, 158–173
　base current, 163
　base impedance, 163
　base power, 158–164, 165–170
　base voltage, 158–164, 165–170
　change of base, 164–165
Permeability, 14
Permeance, 23
Permittivity, 138
Phase
　current, 253

　voltage, 251–253
Phase shift in three-phase transformer banks, 114
Phasor
　definition, 244
　representation of sinusoids, 243–246
Polarity marks for transformers, 94
Potential, 18, 138
Power
　apparent, 251
　average, 248, 249
　complex, 250
　control of, 184–189
　instantaneous, 248
　reactive, 251
　real, 251
　single-phase, 252
　three-phase, 252–254
Power factor, 250, 251
Power flow. *See* Load flow
Power system elements, 3
Primary winding, 93
Propagation constant, 148
Protection, 5
Pullout power, 79

Quasi-static field approximation, 15

Reactive power
　definition, 251
　in synchronous generators, 81
Real power, 251
Regulating transformers, 187–189
Reluctance, 23
Resistance, 126
Resultant MMF, 61
Right-hand rule, 13
Root-mean-square, 249
Rotor, 55, 56, 220, 228
Rotor MMF, 61

Secondary winding, 93
Self-admittance, 175
Short circuit test of transformer, 102, 103
Single-phase motors
　air gap power, 227
　auxiliary winding, 222
　capacitor start motors, 222
　centrifugal switch, 222
　developed power, 228
　equivalent circuit, 223–225

Single-phase motors (*Continued*)
  rotating field, 220, 221
  slip, 225
  split-phase motors, 223
  torque, 221
Sinusoidal steady state
  impedance, 246, 247
  phasor representation, 243–246
Skin effect, 127
Slip, 203
Slip rings, 55
Squirrel cage induction motors, 200
Stall in motors, 205
Stator, 55, 220, 229
Stator MMF, 61
Swing bus, 177
Synchronous machines
  armature reaction, 73
  direct-axis synchronous inductance, 75
  elementary generator, 52
  elementary motor, 193–197
  equivalent circuit, 72, 198
  flux linkages, 56, 59
  generator operation, 52
  motor operation, 197–200
  pullout power, 79
  torque, 59, 67, 197
  torque angle, 78, 199
  voltages, 57, 199
Synchronous speed, 71

Thevenin equivalent circuit for induction motors, 214–216
Three Phase
  Δ connection, 253
  induction motors, 200–219
  line-to-line voltages, 252
  line-to-neutral voltages, 252
  power, 252–254
  transformers, 111
  Y connection, 252
Torque
  dc motors, 235
  induction motors, 210, 211, 217, 221
  synchronous machines, 59, 67
  universal motors, 239
Torque of electric origin, 42, 67
Transformers
  auto, 108, 109
  B-H curve, 110
  connections, 111–115, 117, 118
  cooling, 106
  core losses, 96–98
  core nonlinearities, 109, 110
  current transformation, 95
  efficiency, 106
  equivalent circuits, 103–105
  harmonics, 111
  ideal, 98
  magnetizing current, 96
  multiwinding, 106, 107
  off-nominal turns, 187–189
  open-circuit test, 98
  phase shift, 114
  polarity markings, 94
  regulating, 187–189
  referring quantities across, 99–101
  short-circuit test, 102, 103
  single-phase, 93–110
  three-phase, 116, 117
  voltage regulation, 106
  voltage transformation, 94, 95
Transmission lines
  capacitance, 125, 137, 138
  conductance, 125
  conductor strands, 126, 136
  equivalent circuit, 146–150
  inductance, 125, 127–130
  length classifications, 149, 150
  power flow calculations, 150–152
  resistance, 125, 126
  transposition, 132
Turbines, 48, 49

Universal motors, 238, 239

Voltage
  line-to-line, 252
  line-to-neutral, 252
  phase, 251–253
  in primitive machine, 17–19
  in synchronous generators, 52, 54–59
  transmission line capacitance calculations, 137–139
Voltage control bus, 177
Voltage regulation of transformers, 106

Wound rotor induction motors, 200
Y connections, 252

Y bus matrix, 173–176